AF360924

Micro- and Nano-Fabrication by Metal Assisted Chemical Etching

Micro- and Nano-Fabrication by Metal Assisted Chemical Etching

Editor

Lucia Romano

MDPI • Basel • Beijing • Wuhan • Barcelona • Belgrade • Manchester • Tokyo • Cluj • Tianjin

Editor
Lucia Romano
Paul Scherrer Institut
Switzerland

Editorial Office
MDPI
St. Alban-Anlage 66
4052 Basel, Switzerland

This is a reprint of articles from the Special Issue published online in the open access journal *Micromachines* (ISSN 2072-666X) (available at: https://www.mdpi.com/journal/micromachines/special_issues/Metal_Chemical_Etching).

For citation purposes, cite each article independently as indicated on the article page online and as indicated below:

LastName, A.A.; LastName, B.B.; LastName, C.C. Article Title. *Journal Name* **Year**, *Volume Number*, Page Range.

ISBN 978-3-03943-845-7 (Hbk)
ISBN 978-3-03943-846-4 (PDF)

Cover image courtesy of Lucia Romano.

© 2020 by the authors. Articles in this book are Open Access and distributed under the Creative Commons Attribution (CC BY) license, which allows users to download, copy and build upon published articles, as long as the author and publisher are properly credited, which ensures maximum dissemination and a wider impact of our publications.
The book as a whole is distributed by MDPI under the terms and conditions of the Creative Commons license CC BY-NC-ND.

Contents

About the Editor . . . . **vii**

Lucia Romano
Editorial for the Special Issue on Micro- and Nano-Fabrication by Metal Assisted Chemical Etching
Reprinted from: *Micromachines* **2020**, *11*, 988, doi:10.3390/mi11110988 . . . . **1**

Tae Kyoung Kim, Jee-Hwan Bae, Juyoung Kim, Min Kyung Cho, Yu-Chan Kim, Sungho Jin and Dongwon Chun
Curved Structure of Si by Improving Etching Direction Controllability in Magnetically Guided Metal-Assisted Chemical Etching
Reprinted from: *Micromachines* **2020**, *11*, 744, doi:10.3390/mi11080744 . . . . **3**

Paola Lova and Cesare Soci
Black GaAs: Gold-Assisted Chemical Etching for Light Trapping and Photon Recycling
Reprinted from: *Micromachines* **2020**, *11*, 573, doi:10.3390/mi11060573 . . . . **15**

Oscar Pérez-Díaz and Enrique Quiroga-González
Silicon Conical Structures by Metal Assisted Chemical Etching
Reprinted from: *Micromachines* **2020**, *11*, 402, doi:10.3390/mi11040402 . . . . **27**

Hailiang Li and Changqing Xie
Fabrication of Ultra-High Aspect Ratio (>420:1) Al_2O_3 Nanotube Arraysby Sidewall TransferMetal Assistant Chemical Etching
Reprinted from: *Micromachines* **2020**, *11*, 378, doi:10.3390/mi11040378 . . . . **37**

Thomas Frisk, Fabian Lundberg, Hanna Ohlin, Ulf Johansson, Kenan Li, Anne Sakdinawat and Ulrich Vogt
Metal-Assisted Chemical Etching and Electroless Deposition for Fabrication of Hard X-ray Pd/Si Zone Plates
Reprinted from: *Micromachines* **2020**, *11*, 301, doi:10.3390/mi11030301 . . . . **47**

Olga Volovlikova, Gennady Silakov, Sergey Gavrilov, Tomasz Maniecki and Alexander Dudin
Investigation of the Pd Nanoparticles-Assisted Chemical Etching of Silicon for Ethanol Solution Electrooxidation
Reprinted from: *Micromachines* **2019**, *10*, 872, doi:10.3390/mi10120872 . . . . **57**

Lucia Romano and Marco Stampanoni
Microfabrication of X-ray Optics by Metal Assisted Chemical Etching: A Review
Reprinted from: *Micromachines* **2020**, *11*, 589, doi:10.3390/mi11060589 . . . . **73**

About the Editor

Lucia Romano has a background in materials science for several kinds of applications, from the fundamentals of semiconductors to nanotechnology, from microelectronics to environmental applications and X-ray Optics. Lucia Romano received her Ph.D. in Physics at University of Catania in 2006, and she then did a post-doc at University of Florida. Since 2009, she has been an assistant professor of Physics at the University of Catania, a member of the Ph.D. committee in Materials Science and Nanotechnology, as well as the supervisor of several Master's and Ph.D. students. She was among the scientific staff of several national and international projects, WP leader for the EU project WATER. Since 2014, she has a been a guest professor at ETH-Zurich and a guest scientist at Paul Scherrer Institute, whilst her current research is focused on silicon-based technologies for X-ray optics microfabrication. As an editor and referee for several journals, her track record includes more than 100 publications in international ISI journals (>900 citations) with an h-index of 18.

 *micromachines*

Editorial

Editorial for the Special Issue on Micro- and Nano-Fabrication by Metal Assisted Chemical Etching

Lucia Romano [1,2,3]

[1] Institute for Biomedical Engineering, University and ETH Zürich, 8092 Zürich, Switzerland; lucia.romano@psi.ch

[2] Paul Scherrer Institut, Forschungsstrasse 111, CH-5232 Villigen, Switzerland

[3] Department of Physics and CNR-IMM- University of Catania, 64 via S. Sofia, 95123 Catania, Italy

Received: 31 October 2020; Accepted: 31 October 2020; Published: 31 October 2020

Discovered by Li and Bohn in 2000 [1], Metal Assisted Chemical Etching has been finally recognized as an emerging etching technology. In its twentieth anniversary, Micromachines dedicated this special issue to Metal Assisted Chemical Etching. Several acronyms were reported for this process—metal-assisted chemical etching (MACE), metal-assisted etching (MAE), MacEtch—since 2015 the community seemed to agree with the common acronym of "MacEtch", which was firstly introduced by X. Li [1] to distinguish the unique properties with respect of standard wet-etch and dry-etch techniques. Unlike wet-etch with potassium hydroxide, the MacEtch process is almost independent of crystal orientation and may be used to create a wide variety of patterns, without suffering of micro-loading and scallops of dry-etch. An advantage of the method is the considerable reduction in fabrication costs and complexity with respect to the other techniques. MacEtch is a powerful etching method, capable of fabricating high aspect ratio nano- and micro-structures in few semiconductors substrates (Si, GaAs etc) by using different catalysts (Au, Ag, Pt, Pd etc). Several shapes have been demonstrated with high anisotropy and feature size in the nanoscale: nanoporous film, nanowires, 3D objects, trenches, which are useful components of: photonic devices, microfluidic devices, bio-medical devices, batteries, Vias, microelectromechanical systems (MEMS), X-ray optics etc. With no limitation on large-area and low-cost processing, MacEtch can open up new opportunities for several applications where high precision nano- and micro-fabrication is required. This can make semiconductor manufacturing more accessible to researchers in various fields and accelerates innovation in electronics, bio-medical engineering, energy and photonics.

There are 7 papers published in this Special Issue focusing on several MacEtch applications and fabrication methodologies. Despite the use of metals, which is still the main shortcoming for microelectronics, several manufacturing sectors can benefit of MacEtch. In particular, it seems clear now that the etching capability and the relatively ease of the process are effectively boosting the integration of MacEtch in the microfabrication framework. MacEtch is not only producing porous silicon for new electrochemical applications [2], but also the roughness of GaAs [3] can be tuned for photonic applications. The integration with patterning techniques demonstrated new capabilities from the easy tuning of conical structures [4] to the more complex 3D control by magnetic catalyst [5]. The integration with post-etching processing demonstrated that the etched silicon can functionally serve as template for Al_2O_3 nanotubes [6], Pd zone-plates [7] and Au micro-gratings [8]. While fundamentals are still discussed for GaAs [3], silicon requires fine tuning strategies as a function of the application [2,4–6]. X-ray optics fabrication seems to be the most advanced application of MacEtch with already in use devices [7,8]. Finally, the challenges and the future perspective of MacEtch as nano and micro-fabrication technology are highlighted in the review article [8]. The future looks bright for MacEtch: the recent increment of publications related to MacEtch indicate a gradual integration of the process in microfabrication protocols, with clear advantages of cost-reduction, nano-features capability and high aspect ratio structures. The only recommendation is to be careful in handling and disposing the hazardous hydrofluoric acid!

I would like to take this opportunity to thank all the authors for submitting their papers to this Special Issue. I would also like to thank all the reviewers for dedicating their time and helping to improve the quality of the submitted papers.

Conflicts of Interest: The author declares no conflict of interest.

References

1. Li, X.; Bohn, P.W. Metal-assisted chemical etching in HF/H_2O_2 produces porous silicon. *Appl. Phys. Lett.* **2000**, *77*, 2572–2574. [CrossRef]
2. Volovlikova, O.; Silakov, G.; Gavrilov, S.; Maniecki, T.; Dudin, A. Investigation of the Pd Nanoparticles-Assisted Chemical Etching of Silicon for Ethanol Solution Electrooxidation. *Micromachines* **2019**, *10*, 872. [CrossRef] [PubMed]
3. Lova, P.; Soci, C. Black GaAs: Gold-Assisted Chemical Etching for Light Trapping and Photon Recycling. *Micromachines* **2020**, *11*, 573. [CrossRef] [PubMed]
4. Pérez-Díaz, O.; Quiroga-González, E. Silicon Conical Structures by Metal Assisted Chemical Etching. *Micromachines* **2020**, *11*, 402. [CrossRef] [PubMed]
5. Kim, T.K.; Bae, J.-H.; Kim, J.; Cho, M.K.; Kim, Y.-C.; Jin, S.; Chun, D. Curved Structure of Si by Improving Etching Direction Controllability in Magnetically Guided Metal-Assisted Chemical Etching. *Micromachines* **2020**, *11*, 744. [CrossRef] [PubMed]
6. Li, H.; Xie, C. Fabrication of Ultra-High Aspect Ratio (>420:1) Al_2O_3 Nanotube Arraysby Sidewall TransferMetal Assistant Chemical Etching. *Micromachines* **2020**, *11*, 378. [CrossRef] [PubMed]
7. Akan, R.; Frisk, T.; Lundberg, F.; Ohlin, H.; Johansson, U.; Li, K.; Sakdinawat, A.; Vogt, U. Metal-Assisted Chemical Etching and Electroless Deposition for Fabrication of Hard X-ray Pd/Si Zone Plates. *Micromachines* **2020**, *11*, 301. [CrossRef] [PubMed]
8. Romano, L.; Stampanoni, M. Microfabrication of X-ray Optics by Metal Assisted Chemical Etching: A Review. *Micromachines* **2020**, *11*, 589. [CrossRef]

Publisher's Note: MDPI stays neutral with regard to jurisdictional claims in published maps and institutional affiliations.

© 2020 by the author. Licensee MDPI, Basel, Switzerland. This article is an open access article distributed under the terms and conditions of the Creative Commons Attribution (CC BY) license (http://creativecommons.org/licenses/by/4.0/).

 *micromachines*

Article

Curved Structure of Si by Improving Etching Direction Controllability in Magnetically Guided Metal-Assisted Chemical Etching

Tae Kyoung Kim [1], Jee-Hwan Bae [2], Juyoung Kim [2], Min Kyung Cho [2], Yu-Chan Kim [3], Sungho Jin [1] and Dongwon Chun [2,*

[1] Materials Science and Engineering, University of California at San Diego, 9500 Gilman Dr, La Jolla, CA 92093, USA; tkkim@a123systems.com (T.K.K.); jin1234jin@gamil.com (S.J.)

[2] Advanced Analysis Center, Korea Institute of Science and Technology (KIST), Seoul 02792, Korea; jhbae@kist.re.kr (J.-H.B.); jykim1299@kist.re.kr (J.K.); 023381@kist.re.kr (M.K.C.)

[3] Center for Biomaterials, Biomedical Research Institute, Korea Institute of Science and Technology (KIST), Seoul 02792, Korea; chany@kist.re.kr

* Correspondence: chundream98@kist.re.kr

Received: 1 July 2020; Accepted: 27 July 2020; Published: 30 July 2020

Abstract: Metal-assisted chemical etching (MACE) is widely used to fabricate micro-/nano-structured Si owing to its simplicity and cost-effectiveness. The technique of magnetically guided MACE, involving MACE with a tri-layer metal catalyst, was developed to improve etching speed as well as to adjust the etching direction using an external magnetic field. However, the controllability of the etching direction diminishes with an increase in the etching dimension, owing to the corrosion of Fe due to the etching solution; this impedes the wider application of this approach for the fabrication of complex micro Si structures. In this study, we modified a tri-layer metal catalyst (Au/Fe/Au), wherein the Fe layer was encapsulated to improve direction controllability; this improved controllability was achieved by protecting Fe against the corrosion caused by the etching solution. We demonstrated curved Si microgroove arrays via magnetically guided MACE with Fe encapsulated in the tri-layer catalyst. Furthermore, the curvature in the curved Si microarrays could be modulated via an external magnetic field, indicating that direction controllability could be maintained even for the magnetically guided MACE of bulk Si. The proposed fabrication method developed for producing curved Si microgroove arrays can be applied to electronic devices and micro-electromechanical systems.

Keywords: magnetically guided metal-assisted chemical etching; bulk Si etching; curved Si structure; catalyst encapsulation

1. Introduction

Fabrication techniques for micro/nano-structured Si, such as cantilever beams and bridge and buried microchannels fabricated using micro-electromechanical systems (MEMS) or nano-electromechanical systems (NEMS), have been intensively employed for a wide range of applications, including bio/chemical sensors [1,2], resonators [3], and power generators [4].

Dry-etching methods such as reactive ion etching (RIE) and deep reactive ion etching are important technologies for the micro/nano structuring of Si [5–7]. Recently, the Bosch process, which is a type of RIE, has been applied for the bulk etching of Si with a high aspect ratio, because it enables a high Si etching rate [8]. However, rough scallops on the surface and ion-induced defects impede the wider applications of this technique [9,10]. The deposition of C_4F_8, due to the passivation of the Si sidewall, is unfavorable when reactions are initiated at the Si surface, such as in battery anodes [10–13]. In addition, achieving control over the etching direction in dry-etching methods is difficult, thereby hindering the application of these methods for the fabrication of complex Si structures so that a combined lithography process is required to fabricate MEMS/NEMS structures. An alternative approach is to employ wet Si etching methods using KOH or TMAH (tetramethylammonium hydroxide) etchants to fabricate micro/nano-structured Si; however, the etching direction in such approaches is significantly restricted by the crystalline orientation of Si substrates [14–16].

By contrast, metal-assisted chemical etching (MACE) of Si using noble metal catalysts has garnered substantial attention in the past decade for fabricating micro/nano structures because of its simplicity, low cost, and ability to fabricate high-aspect-ratio structures such as pillar and hole arrays without sidewall etching [17–21]. In general, noble metal catalyst such as Pt, Ag, and Au are used in MACE to significantly improve the etching rate. This is because the Si etching rate in an etchant solution without a catalyst is considerably low because of the low catalytic ability of the Si surface [22]; therefore, only the Si underneath the metal catalyst undergoes considerably etching. Hence, when noble metal catalysts are patterned on the Si surface and immersed in a solution of HF and H_2O_2 diluted in DI water, a faster Si etching rate can be obtained using MACE, as compared to the use of Si without a noble metal catalyst. This implies that the geometry of patterned Si can readily be controlled by the size of the noble metal catalyst, etching time, and etchant concentration in MACE. The MACE of Si has been to fabricate various Si structures such as nanowires and microstructures [23–29].

Ideally, MEMS structures such as Si cantilever and bridge arrays can be obtained via a simple MACE procedure, provided the etching direction can be easily manipulated during the etching process. However, similar to dry/wet etching methods, controlling the etching direction during MACE is difficult, as the etching direction can only be modulated via the crystalline orientation of the Si substrate [30–32] and the concentration of the etchant [33]. To fabricate complex Si structures such as cantilevers and bridges, a type of MEMS structure, the etching direction should be sequentially altered during the etching process. However, the limited control over the etching direction hinders the direct fabrication of the MEMS structure via the MACE of Si. Therefore, a versatile technique for achieving etching direction controllability during MACE is necessary to enable the fabrication of various Si structures that are in demand for industrial applications.

To improve the etching performance in terms of speed and direction control, we developed a magnetically guided MACE process utilizing a magnetic Fe layer encapsulated in a multi-layered metal catalyst [34–36]. Herein, we used a conventional Au/Fe/Au multi-layer metal catalyst instead of a single noble metal; this enabled the adjustment of the etching direction owing to the magnetic pulling force between Fe and an external hard magnet placed underneath the etching bath. The curved Si structures were fabricated by changing the external magnetic field direction during the MACE of Si, the pattern size of which was less than 1 μm [34,36]. Uniform curved Si structures were fabricated via this method; however, as the pattern size in this case exceeded 1 μm, non-uniform curved Si structures were fabricated owing to an inhomogeneous magnetic pulling force, which resulted from the corrosion of Fe due to HF [36]. In addition, the usage of Au-coated magnetic nanoparticles for the curved nanostructure during the magnetically guided MACE of Si is unfavorable if the pattern size needs to be manipulated on a larger scale [34]. Consequently, an appropriate method to suppress the corrosion of Fe due to HF is essential for the fabrication of curved and zig/zag Si microstructures with a massive production yield. This would also eventually facilitate the eventual direct fabrication of MEMS structures such as cantilevers and bridges via MACE.

In this study, we design a novel tri-layer metal catalyst, wherein Fe is encapsulated by Au to suppress its removal by HF, thereby producing a uniform magnetic pulling force. As a result, uniform curved Si structures could be obtained in a bulk scale via magnetically guided MACE. We demonstrate that the tri-layer metal catalyst with encapsulated Fe can suppress Fe corrosion, which, in turn, enables the production of uniformly curved Si structures, as compared to those achieved using a conventional tri-layer metal catalyst.

2. Experimental

Figure 1 summarizes the process of fabricating curved Si structures via magnetically guided MACE. A 500-µm-thick p-type Si (100) wafer (boron-doped; resistivity of 10–20 Ω·cm) was utilized in this study. The Si chip size used for the experiment was 2 × 2 cm². The Si wafer was pre-cleaned in acetone and ethanol for 10 min each. To encapsulate the Fe in the tri-layer Au/Fe/Au catalyst, a photoresist (PR; S1827, MNX, Long Beach, CA, USA; approximately 2.5 µm) line patterning was produced on the Si wafer surface via photolithography (Karl Suss; MA6, SUSS MicroTec, Garching, Germany), as illustrated in Section 3. Subsequently, the Au catalyst layer (20 nm) was deposited on the PR grooves via e-beam evaporation (Temescal BJD 1800 E-beam evaporator, Ferrotec, Santa Clara, CA, USA)). After the PR was removed, the line patterning of Fe was performed to encapsulate Fe with Au, in which the patterning width was narrower than the lower Au layer. Thereafter, the Fe layer (10 nm) was deposited via e-beam evaporation, followed by the removal of the PR. Finally, the PR was patterned such that the pattern width was identical to the lower Au layer. This was also followed by Au deposition (20 nm) via e-beam evaporation to ensure that the Fe layer was protected by the etching solution during the magnetically guide MACE of Si.

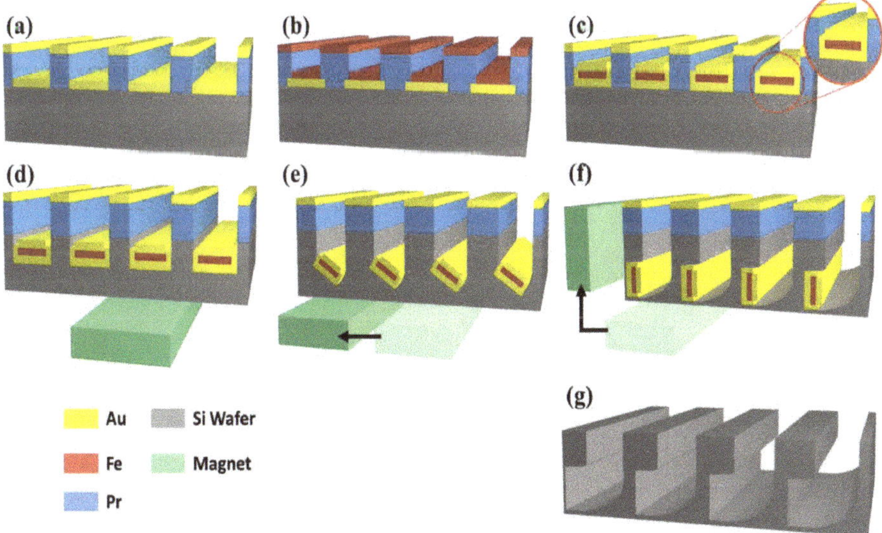

Figure 1. Fabrication process for magnetically guided metal-assisted chemical etching (MACE) of Si: (**a**) photoresist (PR) patterning and bottom Au catalyst (20 nm thickness) deposition, (**b**) PR patterning and magnetic Fe layer (10 nm thickness deposition), (**c**) PR patterning and Au (20 nm thickness) deposition for Fe encapsulation, (**d**), (**e**) and (**f**) magnetically guided MACE of by changing permanent magnet position, (**g**) Si groove arrays after the removal of Au and PR by chemical etching.

After the fabrication of the Au/Fe/Au catalyst encapsulated in Fe via the deposition of an upper Au layer, a post annealing process was performed at 150 °C for 5 h to improve etching performance [37]. High-temperature annealing processes can enhance the ferromagnetic properties of the Fe layer, improving the etching speed and the controllability of the etching direction by increasing the magnetic pulling force. Furthermore, the porous surface structure of the catalyst can be produced via an annealing process that facilitates the diffusion of etchant through the metal catalyst.

A mixture of diluted hydrofluoric acid (2 M) and hydrogen peroxide (9 M) diluted in deionized water was used as an etchant. The magnetically guided MACE of Si was conducted in a Teflon bath; a neodymium iron boron (NdFeB) permanent magnet (4 cm × 4 cm × 1 cm height) with an energy product strength of 35 MGOe was placed outside and underneath this bath to adjust the etching direction during the process. Si chips with the patterned tri-layer Au/Fe/Au catalyst encapsulated in Fe were immersed in the etchant for the magnetically guided MACE. The Si covered by the catalyst was etched faster than that covered by the PR, which can be explained as follows [38].

The reduction of hydrogen peroxide (H_2O_2) on the noble metal catalyst surface generates holes (h^+) as follows:

$$H_2O_2 + 2H^+ \rightarrow 2H_2O + 2h^+ \tag{1}$$

Noble metal catalysts such as Au, Ag, and Pt facilitate the reduction of H_2O_2, thereby improving the etching speed by providing sufficient h^+ than the pure Si surface. The holes generated near the noble metal surface pass through the noble metal surface and are injected into the valence band of Si, because the valance band potential of Si is lower than the redox potential between H_2O and H_2O_2 [22]. This causes the dissolution of Si with HF, as follows:

$$Si + 4h^+ + 6HF \rightarrow SiF_6^{2-} + 6H^+ \tag{2}$$

During the magnetically guided MACE of Si, the permanent magnet was shifted manually to alter the orientation of the magnetic field, which, in turn, changed the etching direction. The Si surface covered by the Au catalyst was etched along the direction of the external magnetic field because of both the enhanced etching rate through the catalyst reaction and the magnetic pulling force between the patterned ferromagnetic Fe and the hard magnet; the area covered by PR was protected because of its resistance to the etchant. The magnetically guided MACE of Si was performed at room temperature; after the etching process was completed, the sample was immersed in acetone and gold etchant to remove the remaining PR and metal catalyst. The etched Si chips were then washed with deionized water. Subsequently, the structure of the tri-layer metal catalyst of Au/Fe/Au and the etched Si structure were characterized using scanning electron microscopy (SEM; Regulus8230, Hitachi, Tokyo, Japan; Phillips XL30 ESEM, FEI, Hillsboro, OR, USA)). In addition, the surface morphology of the tri-layer metal catalyst was characterized via atomic force microscopy (AFM; Park XE7, Park Systems, Suwon, South Korea).

3. Results and Discussion

Figure 2 presents the top-view SEM and energy-dispersive spectroscopy (EDS) images of the tri-layer metal catalyst with encapsulated Fe after the annealing process. Figure 2a,b show the tri-layer metal catalyst with Au (50-μm wide)/Fe (40-μm wide)/Au (47-μm wide) line patterns with 100-μm spacing fabricated on the Si surface via photolithography. This was confirmed by the EDS images of Fe and Au, as shown in Figure 2c; a 45-μm wide Fe layer is patterned between the Au layers. The EDS spectrum at point #2 exhibits sharp peaks at 2.123 KeV and 1.740 KeV corresponding to Au ($M_{\alpha 1}$) and Si ($K_{\alpha 1}$), respectively, whereas only an Si peak was detected at the EDS spectrum of #1, indicating that the lower (50 μm) and upper (47 μm) Au patterns were fabricated successfully. The width of the upper Au layer is narrower than that of the lower Au layer; this discrepancy might be due to a fabrication error. For the EDS spectrum at point #3, Fe $L_{\alpha 1}$ peaks at 0.705 KeV corresponding to Au and Si (Figure 2d) were observed. This indicates that the tri-layer metal catalyst with encapsulated Fe was fabricated successfully.

As the diffusion of the etchant depends on the surface morphology of the metal catalyst [21], porosity is an important factor influencing the etching speed. It has also been reported that enhanced porosity in the metal catalyst results in a higher etching speed [36]. Thus, to improve the porosity of the metal catalyst, we employed an annealing process after the formation of the tri-layer metal catalyst with Fe encapsulation. However, the changes in the surface morphology of the catalyst caused by annealing could not be clearly observed due to limitations of the image resolution of SEM. Hence, AFM was employed to analyze the surface morphologies before (Figure 3a) and after (Figure 3b) the annealing process. The root-mean-square roughness of the annealed metal catalyst was 1.038 nm (Figure 3b), whereas that of the as-deposited metal catalyst was 0.532 nm; this indicates that annealing provides a rough surface morphology that favors a higher etching during the magnetically guided MACE of Si, due to the shorter diffusion length of the etching solution [21].

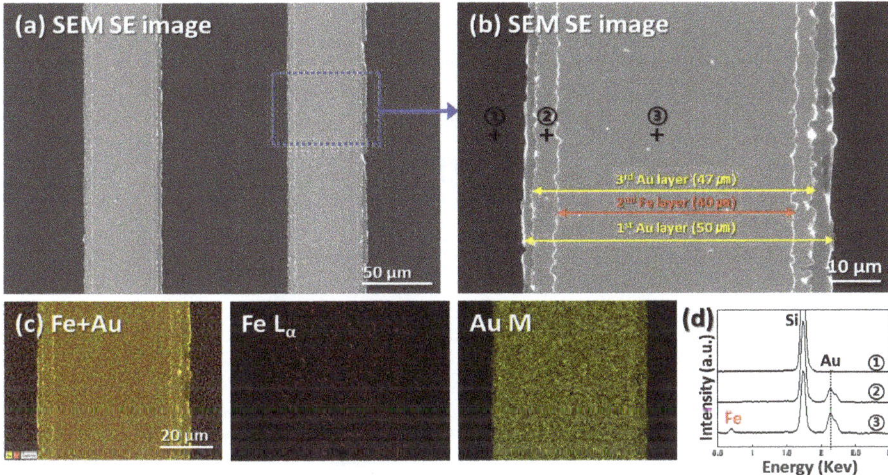

Figure 2. Scanning electron microscopy (SEM) images and energy-dispersive spectroscopy (EDS) images and spectrums of the tri-layer Au/Fe/Au metal catalyst encapsulating Fe after the annealing process.

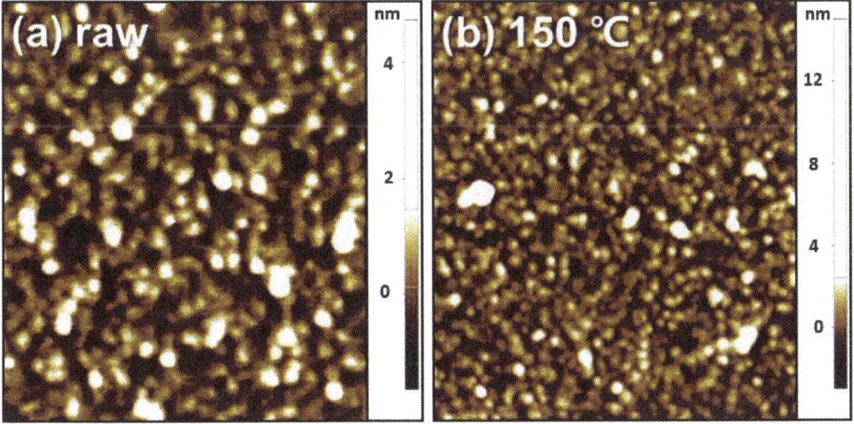

Figure 3. Atomic force microscopy (AFM) images of the tri-layer Au/Fe/Au metal catalyst with Fe encapsulation before (**a**) and after (**b**) the annealing process.

Figure 4 shows the cross-sectional SEM images of vertical groove arrays produced via the magnetically guided MACE of Si for 3 h (Figure 4a) and 5 h (Figure 4b). The tri-layer metal catalyst of Au (20-µm wide)/Fe (15-µm wide)/Au (20-µm wide) lines spaced at 100 µm intervals were patterned on the surface of Si encapsulated with Fe, and the MACE of Si was performed under the magnetic field. During the etching, the strong magnet was placed underneath the Teflon beaker, and its position remained fixed for vertical etching.

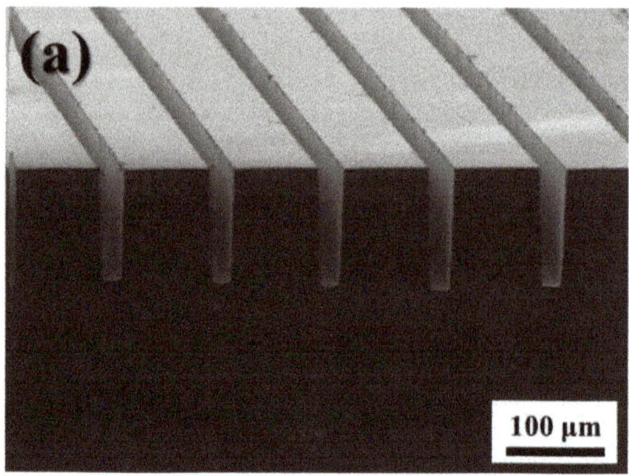

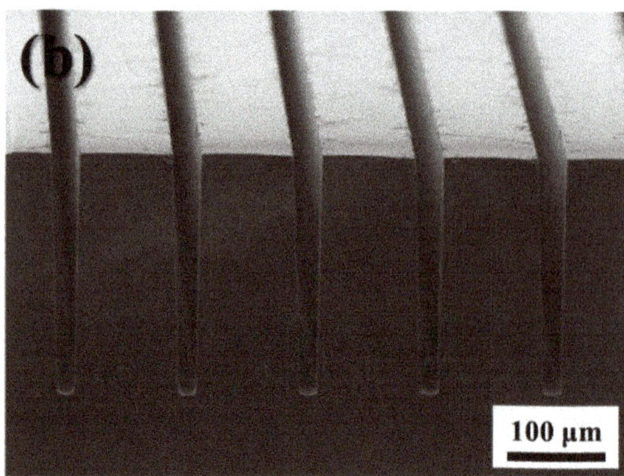

Figure 4. Cross-sectional SEM images of vertical Si groove arrays produced by magnetically guided MACE for etching durations of (**a**) 3 h and (**b**) 5 h.

Figure 4 confirms that the Au/Fe/Au tri-layer metal catalyst with encapsulated Fe is suitable for the magnetically guided MACE of Si in the vertical direction. The measured groove thickness for the etching of 3 h was 150 μm, while that for the etching of 5 h was 240 μm; these correspond to etching speeds of 0.83 μm/min (Figure 4a) and 0.8 μm/min (Figure 4b), respectively. Likewise, a reduction in the etching speed with etching time was observed in our previous results, which were obtained using a conventional Au/Fe/Au tri-layer metal catalyst [34,35]. A direct comparison between etching speed and catalyst structure is difficult because etching speed depends on not only the catalyst structure but etchant concentration and the size of the catalyst patterned on the Si surface during magnetically guided MACE. However, the reduction in etching speed with increasing time appears to be mitigated by the encapsulation of Fe, as compared to the etching speed for a conventional Au/Fe/Au catalyst [34,35]. This reduction in etching speed with the increase in etching time can be attributed to the reduction in the volume of Fe caused by the HF-induced corrosion, which decreases the magnetic pulling force between Fe and the permanent magnet. Unlike the conventional Au/Fe/Au tri-layer catalyst, the sidewalls of the Fe layer in our catalyst remain protected due to the encapsulation by Au; this effectively suppresses Fe corrosion. As a result, the reduction in etching speed with an increase in etching time is not appreciable during the magnetically guided MACE of Si. It should be noted that a slightly tapered structure was produced. The loss at both Au layers in the metal catalyst contribute toward the corrosion of Fe, resulting in a reduction of catalyst dimensions with increasing etching time; this could potentially explain the tapered Si structure obtained. Moreover, this indicates that the corrosion of Fe could not be prevented completely, even when Fe is encapsulated by Au.

Figure 5 shows the Si groove arrays produced by magnetically guided MACE for 5 h using a conventional tri-layer catalyst (Figure 5a) and the tri-layer catalyst with encapsulated Fe (Figure 5b). The tri-layer metal catalyst of Au (50 µm wide)/Fe (40 µm wide)/Au (50 µm wide) line patterns were produced on the surface of Si encapsulated by Fe. To adjust the etching direction for fabricating a curved Si structure via magnetically guided MACE, the magnet underneath the beaker was moved gradually from a vertical position to a tilting angle of 30° during the etching.

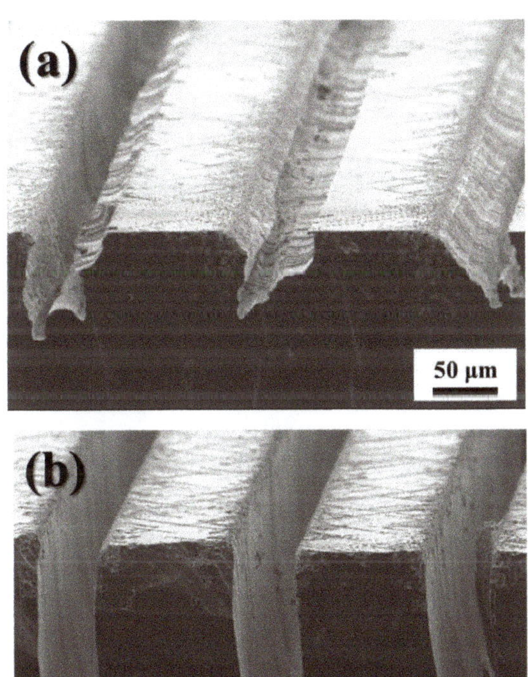

Figure 5. Curved Si groove arrays produced by magnetically guided MACE for 5 h using (**a**) a Swiss-cheese-like Au/Fe/Au catalyst and (**b**) Au/Fe$_{encapsulated}$/Au catalyst (tilting angle: 30°).

As shown in Figure 5a, 60-µm-thick and non-uniform Si grooves with rough surfaces were produced by etching with a conventional tri-layer catalyst, although rougher surface morphology was produced by annealing process, as shown in Figure 3, which reduces the diffusion length of the etchant. Changing the etching direction results in an unstable Si/catalyst interface, thereby facilitating Fe corrosion because the etchant can readily penetrate the interface and etch the Fe. Therefore, as the Fe is not sufficiently protected given the changing direction of the magnetic field, magnetically guided MACE of Si is infeasible due to the reduced and inhomogeneous magnetic pulling force. In contrast, uniform 150-µm-wide curved Si groove arrays with smooth surfaces were produced during the magnetically guided MACE of Si using the tri-layer metal catalyst with encapsulated Fe, as shown in Figure 6b; this indicates that the etching direction can be effectively controlled by changing the direction of the magnetic field when the catalyst includes encapsulated Fe, thereby proving that the HF-induced corrosion of Fe can be effectively suppressed via encapsulation.

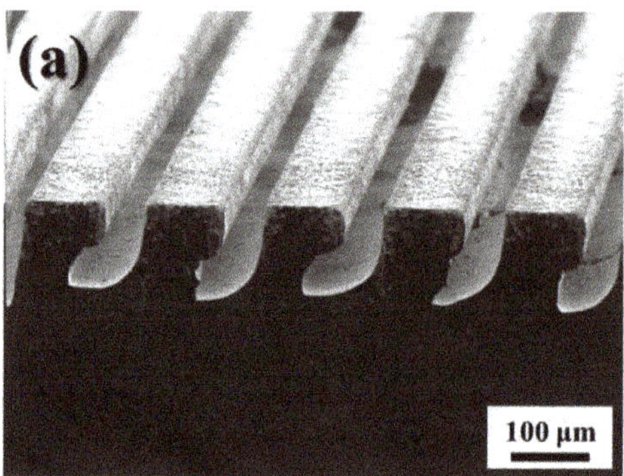

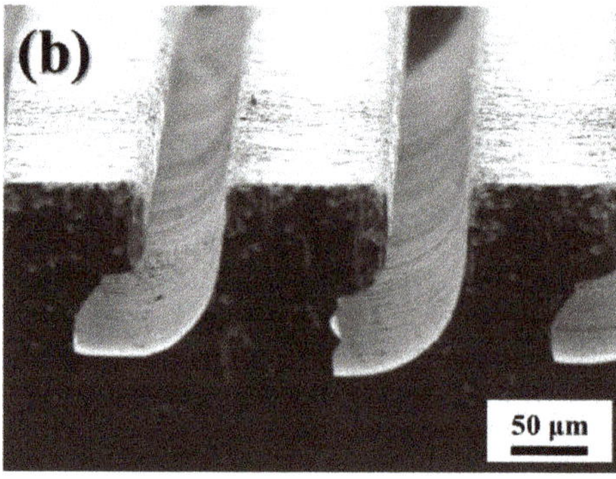

Figure 6. Curved Si groove arrays produced by magnetically guided MACE for 5 h using (**a**) a Swiss-cheese-like Au/Fe/Au catalyst and (**b**) Au / Fe$_{encapsulated}$ /Au catalyst (tilting angle: 90°).

The etching speeds calculated from the thickness of the measured grooves in Figures 5a and 4b are approximately 2 μm/min and 0.5 μm/min, respectively. Fe corrosion could result in partial Au ejection because of the strong adhesive force between Fe/Au, as compared with that for Au/Si. This would result in a lower etching speed and a rougher surface when using magnetically guided MACE of Si with a conventional tri-layer metal catalyst, as shown in Figure 6a. In contrast, a higher etching speed and a smooth surface morphology were obtained when the tri-layer metal catalyst with encapsulated Fe was used, indicating the partial Au ejection was successfully prevented by the suppression of Fe corrosion.

Figure 6 presents the curved Si groove arrays produced by magnetically guided MACE for 5 h using the tri-layer metal catalyst with encapsulated Fe. The tri-layer metal catalyst of Au (50-μm wide)/Fe (40-μm wide)/Au (50-μm wide) line patterns spaced at 100 μm intervals were fabricated on the surface of Si. First, the magnetically guided MACE of Si was performed by using a strong magnet placed vertically under a beaker. After this vertical etching for 1 h, the position of the hard magnet was gradually changed to a 90° tilting angle, and etching was conducted for 2 h under a lateral magnetic field.

As shown in Figure 6, a curved Si structure with a smooth surface was produced via magnetically guided MACE when the tri-layer metal catalyst with encapsulated Fe was used. As expected, the curvature of the Si structure was enhanced on increasing the tilting angle from 30° to 90°, indicating that the etching direction could be adjusted effectively by using the external magnetic field.

Finally, the bulk micromachining of Si with a curved structure was successful conducted via the magnetically guided MACE of Si using the metal catalyst with Fe encapsulated in an Au (50 μm wide)/Fe (40 μm wide)/Au (50 μm wide) tri-layer. Therefore, it is evident that suppressing the corrosion of Fe, through Fe encapsulation, improves the controllability of etching direction during the magnetically guided MACE of Si.

4. Conclusions

Uniform curved Si groove arrays were produced using encapsulated Fe in a tri-layer metal catalyst (Au/Fe/Au); this catalyst improved the controllability of etching direction during magnetically guided MACE. A magnetic layer in a tri-layer metal catalyst was used to modulate the etching direction via a magnetic pulling force. To improve the direction controllability during magnetically guided MACE, the Fe layer was encapsulated with Au, which suppresses the HF-induced corrosion of Fe, thereby enabling the production of a uniform magnetic pulling force.

Uniform curved Si groove arrays were produced via magnetically guided MACE using this Fe encapsulation, whereby the etching direction was manipulated by moving the external strong magnet during this process. By contrast, non-uniform Si groove arrays with rougher surfaces were obtained when using a conventional tri-layer metal catalyst. This indicates that the encapsulation of Fe with Au can effectively protect Fe against corrosion due to HF, resulting in an enhanced direction controllability. Furthermore, it was observed that the curvature of Si groove arrays can be modulated by adjusting the tilting angle of the external strong magnet, when using encapsulated Fe in the tri-layer metal catalyst.

We also attempted to fabricate complex Si structures such as Si cantilever arrays and zig-zag microwire arrays; however, this could not be achieved using the proposed approach. On increasing the etching time, the controllability of etching direction decreases, which is likely caused by the corrosion of Fe due to HF. In addition, depletion of the etchant around etched areas between the metal catalyst and Si could result in a reduced etching rate as well as poor controllability over etching direction. To directly fabricate MEMS structures via the magnetically guided MACE of Si, magnetic alloy materials such as AuFe and CoFe need to be used as the metal catalyst in order to reduce the corrosion of the magnetic layer by HF. In addition, the etching process should be conducted under an etchant circulation system to ensure sufficient etchant reaches the reactive area which remain as future works. Although complex Si structures could not be achieved via magnetically guided MACE with Fe encapsulation, the results of this study prove that encapsulation helps prevent Fe corrosion, resulting in a uniform magnetic pulling force. Consequently, uniform curved Si structures can be fabricated in bulk by improving the controllability of etching direction during magnetically guided MACE.

Author Contributions: T.K.K., Y.-C.K., S.J., and D.C. designed the experiments; T.K.K., J.-H.B., and D.C. wrote the manuscript. T.K.K. and M.K.C. prepared Figures 2–6; J.-H.B. and J.K. prepared Figure 1, Figure 2, and Figure 6. All authors have read and agreed to the published version of the manuscript.

Funding: This work was supported by the National Research Foundation (NRF) of Korea under Grant 2018M3D1A1058793, funded by the government of Korea.

Conflicts of Interest: The authors declare no conflict of interest.

References

1. Mader, A.; Gruber, K.; Castelli, R.; Hermann, B.A.; Seeberger, P.H.; Rädler, J.O.; Leisner, M. Discrimination of Escherichia coli Strains using Glycan Cantilever Array Sensors. *Nano Lett.* **2012**, *12*, 420–423. [CrossRef]
2. Hunter, G.W.; Akbar, S.; Bhansali, S.; Daniele, M.; Erb, P.D.; Johnson, K.; Liu, C.-C.; Miller, D.; Oralkan, O.; Hesketh, P.J. A Critical Review of Solid State Gas Sensors. *J. Electrochem. Soc.* **2020**, *167*, 037570. [CrossRef]
3. Tao, Y.; Navaretti, P.; Hauert, R.; Grob, U.; Poggio, M.; Degen, C.L. Permanent reduction of dissipation in nanomechanical Si resonators by chemical surface protection. *Nanotechnology* **2015**, *26*, 465501. [CrossRef] [PubMed]

4.   Shen, D.; Park, J.-H.; Ajitsaria, J.; Choe, S.-Y.; Wikle, H.C., III; Kim, D.-J. The design, fabrication and evaluation of a MEMS PZT cantilever with an integrated Si proof mass for vibration energy harvesting. *J. Micromech. Microeng.* **2008**, *18*, 055017. [CrossRef]

5.   Gosálves, M.A.; Zubel, I.; Viinikka, E. *Wet Etching of Silicon, Handbook of Silicon Based Mems Materials and Technologies*; William Andrew Publishing: New York, NY, USA, 2010.

6.   Stöhr, F.; Michael-Lindhard, J.; Hübner, J.; Jensen, F.; Simons, H.; Jakobsen, A.C.; Poulsen, H.F.; Hansen, O. Sacrificial structures for deep reactive ion etching of high-aspect ratio kinoform silicon x-ray lenses. *J. Vac. Sci. Technol. B Nanotechnol. Microelectron.* **2015**, *33*, 062001. [CrossRef]

7.   Tang, Y.; Sandoughsaz, A.; Owen, K.J.; Najafi, K. Ultra deep reactive ion etching of high aspect-ratio and thick silicon using a ramped-parameter process. *J. Microelectromech. Syst.* **2018**, *27*, 686–697. [CrossRef]

8.   Toan, N.V.; Toda, M.; Kawai, Y.; Ono, T. A capacitive silicon resonator with a movable electrode structure for gap width reduction. *J. Micromech. Microeng.* **2014**, *24*, 025006. [CrossRef]

9.   Toan, N.V.; Kubota, T.; Sekhar, H.; Samukawa, S.; Ono, T. Mechanical quality factor enhancement in a silicon micromechanical resonator by low-damage process using neutral beam etching technology. *J. Micromech. Microeng.* **2014**, *24*, 085005. [CrossRef]

10.   Fu, J.; Li, J.; Yu, J.; Liu, R.; Li, J.; Wang, W.; Wang, W.; Chen, D. Improving sidewall roughness by combined RIE-Bosch process. *Mat. Sci. Semicon. Proc.* **2018**, *83*, 186–191. [CrossRef]

11.   Jo, S.-B.; Lee, M.-W.; Lee, S.-G.; Lee, E.-H.; Park, S.-G.; O, B.-H. Characterization of a modified Bosch-type process for silicon mold fabrication. *JVST A* **2005**, *23*, 905–910. [CrossRef]

12.   Bang, B.M.; Kim, H.; Song, H.-K.; Cho, J.; Park, S. Scalable approach to multi-dimensional bulk Si anodes via metal-assisted chemical etching. *Energy Environ. Sci.* **2011**, *4*, 5013–5019. [CrossRef]

13.   Chan, C.K.; Peng, H.; Liu, G.; McIlwrath, K.; Zhang, X.F.; Huggins, R.A.; Cui, Y. High-Performance lithium battery anodes using silicon nanowires. *Nat. Nanotechnol.* **2008**, *3*, 31–35. [CrossRef] [PubMed]

14.   Kramkowska, M.; Zubel, I. Silicon anisotropic etching in KOH and TMAH with modified surface tension. *Procedia Chem.* **2009**, *1*, 774–777. [CrossRef]

15.   Pal, P.; Sato, K. A comprehensive review on convex and concave corners in silicon bulk micromachining based on anisotropic wet chemical etching. *Micro Nano Syst. Lett.* **2015**, *3*, 6. [CrossRef]

16.   Madou, M.J. *Fundamentals of Microfabrication and Nanotechnology*; Three-volume set; CRC Press: Florida, FL, USA, 2018.

17.   Li, X. Metal assisted chemical etching for high aspect ratio nanostructures: A review of characteristics and applications in photovoltaics. *Curr. Opin. Solid State Mater. Sci.* **2012**, *16*, 71–81. [CrossRef]

18.   Peng, K.; Lu, A.; Zhang, R.; Lee, S.T. Motility of metal nanoparticles in silicon and induced anisotropic silicon etching. *Adv. Funct. Mater.* **2008**, *18*, 3026–3035. [CrossRef]

19.   Kong, L.; Zhao, Y.; Dasgupta, B.; Ren, Y.; Hippalgaonkar, K.; Li, X.; Chim, W.K.; Chiam, S.Y. Minimizing Isolate Catalyst Motion in Metal-Assisted Chemical Etching for Deep Trenching of Silicon Nanohole Array. *ACS Appl. Mater. Interfaces* **2017**, *9*, 20981–20990. [CrossRef]

20.   Wu, B.; Kumar, A.; Pamarthy, S. High aspect ratio silicon etch: A review. *J. Appl. Phys.* **2010**, *108*, 9. [CrossRef]

21.   Um, H.-D.; Kim, N.; Lee, K.; Hwang, I.; Seo, J.H.; Young, J.Y.; Duane, P.; Wober, M.; Seo, K. Versatile control of metal-assisted chemical etching for vertical silicon microwire arrays and their photovoltaic applications. *Sci. Rep.* **2015**, *5*, 11277. [CrossRef]

22.   Huang, Z.; Geyer, N.; Werner, P.; De Boor, J.; Gösele, U. Metal-Assisted Chemical Etching of Silicon: A Review: In memory of Prof. *Ulrich Gösele, Adv. Mater.* **2011**, *23*, 285–308. [CrossRef]

23.   Choi, W.; Liew, T.; Dawood, M.; Smith, H.I.; Thompson, C.; Hong, M. Synthesis of silicon nanowires and nanofin arrays using interference lithography and catalytic etching. *Nano Lett.* **2008**, *8*, 3799–3802. [CrossRef] [PubMed]

24.   Huang, Z.; Zhang, X.; Reiche, M.; Liu, L.; Lee, W.; Shimizu, T.; Senz, S.; Gösele, U. Extended arrays of vertically aligned sub-10 nm diameter [100] Si nanowires by metal-assisted chemical etching. *Nano Lett.* **2008**, *8*, 3046–3051. [CrossRef] [PubMed]

25.   Choi, H.-J.; Baek, S.; Jang, H.S.; Kim, S.B.; Oh, B.-Y.; Kim, J.H. Optimization of metal-assisted chemical etching process in fabrication of p-type silicon wire arrays. *Curr. Appl. Phys.* **2011**, *11*, S25–S29. [CrossRef]

26.   Zahedinejad, M.; Farimani, S.D.; Khaje, M.; Mehrara, H.; Erfanian, A.; Zeinali, F. Deep and vertical silicon bulk micromachining using metal assisted chemical etching. *J. Micromech. Microeng.* **2013**, *23*, 055015. [CrossRef]

Micromachines **2020**, *11*, 744

27. Lee, M.-H.; Khang, D.-Y. Facile generation of surface structures having opposite tone in metal-assisted chemical etching of Si: Pillars vs. holes. *RSC Adv.* **2013**, *3*, 26313–26320. [CrossRef]

28. Kim, S.M.; Khang, D.Y. Bulk Micromachining of Si by Metal-assisted Chemical Etching. *Small* **2014**, *10*, 3761–3766. [CrossRef]

29. Bassu, M.; Surdo, S.; Strambini, L.M.; Barillaro, G. Electrochemical micromachining as an enabling technology for advanced silicon microstructuring. *Adv. Funct. Mater.* **2012**, *22*, 1222–1228. [CrossRef]

30. Peng, K.; Wu, Y.; Fang, H.; Zhong, X.; Xu, Y.; Zhu, J. Uniform, axial-orientation alignment of one-dimensional single-crystal silicon nanostructure arrays. *Angew. Chem. Int. Ed. Engl.* **2005**, *44*, 2737–2742. [CrossRef]

31. Kim, J.; Kim, Y.H.; Choi, S.-H.; Lee, W. Curved silicon nanowires with ribbon-like cross sections by metal-assisted chemical etching. *ACS Nano* **2011**, *5*, 5242–5248. [CrossRef]

32. Kolasinski, K.W.; Unger, B.A.; Ernst, A.T.; Aindow, M. Crystallographically Determined Etching and Its Relevance to the Metal-Assisted Catalytic Etching (MACE) of Silicon Powders. *Front. Chem.* **2018**, *6*, 651. [CrossRef]

33. Han, H.; Huang, Z.; Lee, W. Metal-Assisted chemical etching of silicon and nanotechnology applications. *Nano Today* **2014**, *9*, 271–304. [CrossRef]

34. Oh, Y.; Choi, C.; Hong, D.; Kong, S.D.; Jin, S. Magnetically Guided Nano-Micro Shaping and Slicing of Silicon. *Nano Lett.* **2012**, *12*, 2045–2050. [CrossRef]

35. Chun, D.W.; Kim, T.K.; Choi, D.; Caldwell, E.; Kim, Y.J.; Paik, J.C.; Jin, S.; Chen, R.K. Vertical Si nanowire arrays fabricated by magnetically guided metal-assisted chemical etching. *Nanotechnology* **2016**, *27*. [CrossRef] [PubMed]

36. Kim, T.K.; Bae, J.-H.; Kim, J.; Kim, Y.-C.; Jin, S.; Chun, D.W. Bulk Micromachining of Si by Annealing-Driven Magnetically Guided Metal-Assisted Chemical Etching. *ACS Appl. Electron. Mater.* **2020**, *2*, 260–267. [CrossRef]

37. Borgohain, D.; Dash, R.K. Understanding the influence of thermal annealing of the metal catalyst on the metal assisted chemical etching of silicon. *J. Mater. Sci.-Mater. El.* **2018**, *29*, 4211–4216. [CrossRef]

38. Chattopadhyay, S.; Li, X.; Bohn, P.W. In-Plane control of morphology and tunable photoluminescence in porous silicon produced by metal-assisted electroless chemical etching. *J. Appl. Phys.* **2002**, *91*, 6134–6140. [CrossRef]

© 2020 by the authors. Licensee MDPI, Basel, Switzerland. This article is an open access article distributed under the terms and conditions of the Creative Commons Attribution (CC BY) license (http://creativecommons.org/licenses/by/4.0/).

 *micromachines*

Article

Black GaAs: Gold-Assisted Chemical Etching for Light Trapping and Photon Recycling

Paola Lova *,† and Cesare Soci

School of Physical and Mathematical Sciences, Division of Physics and Applied Physics,
Nanyang Technological University, 21 Nanyang Link, Singapore 637371, Singapore; csoci@ntu.edu.sg
* Correspondence: lova.paola@unige.it; Tel.: +39-010-353-6192
† Current Address: Dipartimento di Chimica e Chimica Industriale, Università degli Studi di Genova, Via Dodecaneso 31, 16146 Genova, Italy.

Received: 17 April 2020; Accepted: 4 June 2020; Published: 5 June 2020

Abstract: Thanks to its excellent semiconductor properties, like high charge carrier mobility and absorption coefficient in the near infrared spectral region, GaAs is the material of choice for thin film photovoltaic devices. Because of its high reflectivity, surface microstructuring is a viable approach to further enhance photon absorption of GaAs and improve photovoltaic performance. To this end, metal-assisted chemical etching represents a simple, low-cost, and easy to scale-up microstructuring method, particularly when compared to dry etching methods. In this work, we show that the etched GaAs (*black GaAs*) has exceptional light trapping properties inducing a 120 times lower surface reflectance than that of polished GaAs and that the structured surface favors photon recycling. As a proof of principle, we investigate photon reabsorption in hybrid GaAs:poly (3-hexylthiophene) heterointerfaces.

Keywords: metal-assisted chemical etching; antireflection; *black GaAs*; photon recycling

1. Introduction

In the last decades several antireflective coatings, scattering layers and photon recycling structures have been developed to enhance photovoltaic performances in devices based on inorganic semiconductors, [1–3] donor-acceptor organic [4] and hybrid [5,6] systems, perovskites [7], and for light transmitting panels based on luminescent solar concentrators [8]. Photon absorption can also be enhanced using scattering and light trapping surfaces [9–12] or photonic crystals [13–15] that aim to increase the light optical path, to confine light in the active layer, or to recycle emitted photons [16]. Most of these structures are obtained by lithographic methods that produce high quality and homogeneous patterns, but often do not provide the high throughput necessary for the fabrication of large area devices. To this end, wet chemical methods are a viable, low-cost, and easy to scale-up alternative to lithographic patterning. Among these processes, metal-assisted chemical etching (MacEtch) became a paradigm for silicon structuring to produce surfaces with near-zero reflectance [17–25]. Etched black silicon is indeed a commercial product, and entered mass production of solar cells and modules [26,27]. The high throughput demonstrated for silicon made wet etching processes interesting also for translation to other semiconductors [28].

MacEtch relies on the dissolution of a semiconductor surface catalyzed by a metal in a bath containing alkaline or acidic oxidizing agents [17,20,23,29–31]. The catalyst favors anisotropic substrate dissolution. Then, the substrate can be structured by patterned metal films [32,33] or etched randomly using dispersed metallic particles [24]. In recent years, several attempts aimed to extend MacEtch to III-V group semiconductors [32–43], which yield better device characteristics in light emitting diodes and solar cells compared to mainstream silicon and germanium [44,45]. GaAs structuring via MacEtch has been reported in conjunction with catalyst vacuum depositions [46], or with metal patterning [32–41] by nanoimprint lithography [47], photolithography [20], and microsphere self-assembly [48]. While lithographic depositions allow highly controlled nanostructuring, they have limited room for scaling-up the fabrication. Recently, lithography-free MacEtch of GaAs was demonstrated by both electrode [49] and electrodeless deposition of gold nanoparticles [12]. In the first case, etching the (100) GaAs surface in a $KMnO_4$:HF bath yields nanowire arrays similar to those reported for etched black silicon. This approach generates an effective medium with reflectance as low as 4% [49]. In the second case, etching in H_2O_2/HF baths induces crystal plane dependent dissolution rates and then light trapping surfaces with reflectance lower than 2%. [12] In this case, the crystal plane dependent etching rate is favored by the different reactivity of the Ga and As sp_3 atoms [12]. A schematic of the etching reactions is illustrated in Figure 1. Specifically, a gold nanoparticle previously cast on the substrate is oxidized by the peroxide in the bath. The oxidized gold cations diffuse along the semiconductor surface and are selectively reduced at an arsenic site. Then, the precipitated catalyst can be further oxidized or form a electrochemical cell for gold deposition and surface etching [12,50]. Although catalyst ions diffusion and precipitation mechanism apparently contradict the etching mechanism reported in the seminal paper by Li and Bohn [17] and further confirmed by a number of researches [32], catalyst dissolution and redeposition have been demonstrated for two-step etching processes [50,51]. The selective auric reduction is favored by the higher reactivity of As with respect to Ga in the structure [52]. Indeed, the zinc-blende structure stabilizes trivalent Ga, and destabilizes As, which has valence five in electrophilic environment and an easy-to-oxidize pair of electrons. Arsenic is then oxidized to arsenic and arsenious acids, while Ga is complexed by fluorine anions allowing migration of the catalyst within the inner surface. Normally, monoatomic Ga planes, which only expose Ga atoms, are mildly reactive in the etchant solution. However, the catalytic effect of migrating gold ions is inhibited when a continuous catalyst film is cast on the semiconductor surface. In this case, the etching process does not depend on the crystal plane exposed to the bath [33,53–57].

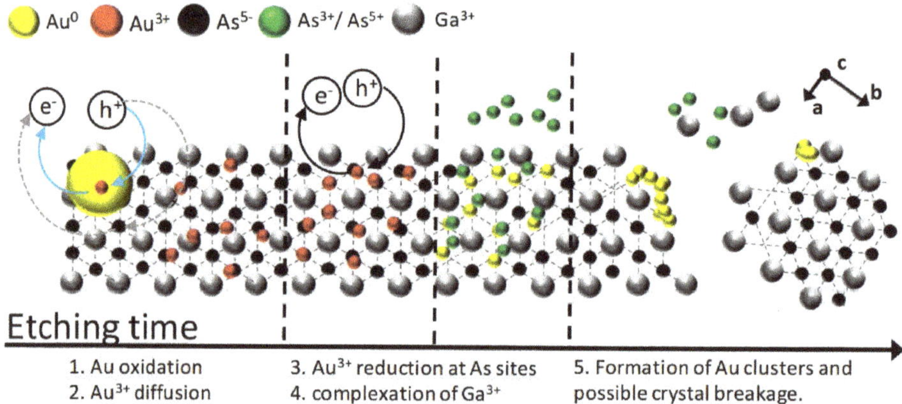

Figure 1. Schematic of the metal-assisted chemical etching (MacEtch) process of a GaAs (111)B surface.

As mentioned above, while lithographic patterning limits the scalability and throughput of the etching process, lithography-free MacEtch is a start-to-end solution process that is easy to scale up to wafer size. The etching of different GaAs crystallographic orientations was previously reported and light trapping properties were reported for etched GaAs (111)B and (100) surfaces [12]. In this work, we investigate the possibility to employ the etched structures for photon recycling in hybrid polymer-inorganic heterointerfaces. Reabsorption of emitted photons is indeed particularly desirable in hybrid photovoltaic devices which suffer of low power conversion efficiency related to high charges recombination and low charge carrier mobility [58–60]. The possibility to reabsorb photons emitted upon charge recombination, which would otherwise increase losses, could improve the performance of these devices [58–60]. It was indeed demonstrated that charge generation at hybrid heterointerfaces occurs in both the polymer and in the inorganic material [6]. Photon reabsorption, therefore, could lead to larger generation yield in hybrid devices. To assess the possibility to employ the process for the fabrication of photon recycling surfaces, as a proof of principle, we investigate the photoluminescence reabsorption occurring at the heterointerface between structured GaAs and poly(3-hexiltiphene) (P3HT), which is a widely studied conjugated polymer in hybrid and tandem photovoltaic devices.

2. Materials and Methods

Metal-assisted chemical etching: the MacEtch was performed at room temperature on *n*-type (100), (110), (111)B, and (211) epi-ready GaAs wafers (Axt Inc, Fremont, CA, USA). In the process, gold nanoparticles were first deposited on the GaAs surfaces by immersion in a water solution containing 0.1 mM of $AuCl_3$ (Sigma Aldrich, Saint Louis, MO, USA). The samples were successively blow-dried and etched in a bath of HF and H_2O_2 (4:1) in a Teflon® container (DuPont Wilmington, DE, United States) for about 10 min under vigorous stirring. The samples were then removed from the bath, rinsed in deionized water, and finally blow-dried. Au nanoparticles removal was performed by further etching of the (111)B microstructured surface in aqua regia (HNO_3/HCl = 1:3) with different dilution in water (not diluted, 1:5, and 1:10).

P3HT deposition: P3HT was deposited on quartz substrates (Hellma Müllheim, Baden Württemberg, Germany) and on polished and etched GaAs by spin coating a 10% (*w/v*) solution in dichlorobenzene at a rotation speed of 800 rpm. All the samples were annealed for 15 min at 120 °C on a hotplate (VWR, Radnor, PA, United States). Deposition and annealing were performed in nitrogen environment to avoid polymer oxidation.

Structural and optical characterization: Gold nanoparticles and microstructure images of the etched GaAs samples were collected by scanning electron microscopy (SEM) with a field emission Jeol JSM-6700F (Jeol, Akishima, Japan) endowed with a secondary electron detector. Particle size distribution was retrieved manually. Representations of the GaAs crystal lattice were elaborated with the software VESTA (JP-Minerals, Ibaraki, Japan) [61] using data retrieved from the Crystallography Open Database [62]. Normal incidence reflectance was collected with a Bruker Vertex 80v Fourier transform spectrometer (Bruker, Billica, MA, USA) coupled with a Bruker HYPERION microscope (Bruker). Photoluminescence spectra were collected at room temperature using a Horiba Fluorolog spectrofluorometer (Horiba, Kyoto, Japan) with a CCD detector, a Xenon lamp as the exciting source, and a monochromator to select excitation wavelength. The measurements were collected exciting both P3HT and GaAs at 500 nm.

3. Results and Discussion

The effective electrodeless deposition of Au nanoparticles on the GaAs surface was confirmed by SEM measurements. Figure 2a reports a micrograph of the pristine particles, which are distributed densely and homogeneously on the GaAs surface. The diameter distribution shown in Figure 2c in black color displays that the catalyst has a broad size dispersion, whit diameters ranging from 3 nm to about 50 nm, while the larger fraction has a diameter approaching 7 nm. These data suggest the presence of particles aggregates on the semiconductor surface, which are not easily recognizable form single particles owing to the instrumental resolution. Figure 2b reports instead a SEM micrograph showing the gold particles on the surface after the etching process. In this case, they are not homogeneously distributed over the sample. Indeed after the etching, the catalysts are located within the etched features (see Figures 3 and 4). However, the particles dimension is more homogenous with respect to the previous case. The size dispersion shown in red in Figure 2c indicates a sharper distribution with diameters ranging from 6 to 35 nm, with the larger percentage having a diameter equal to 11 nm. The variations in the particles size and distribution can be attributed to the continuous gold oxidation and reduction, which implies its dissolution and reprecipitation may decrease the diameter polydispersity.

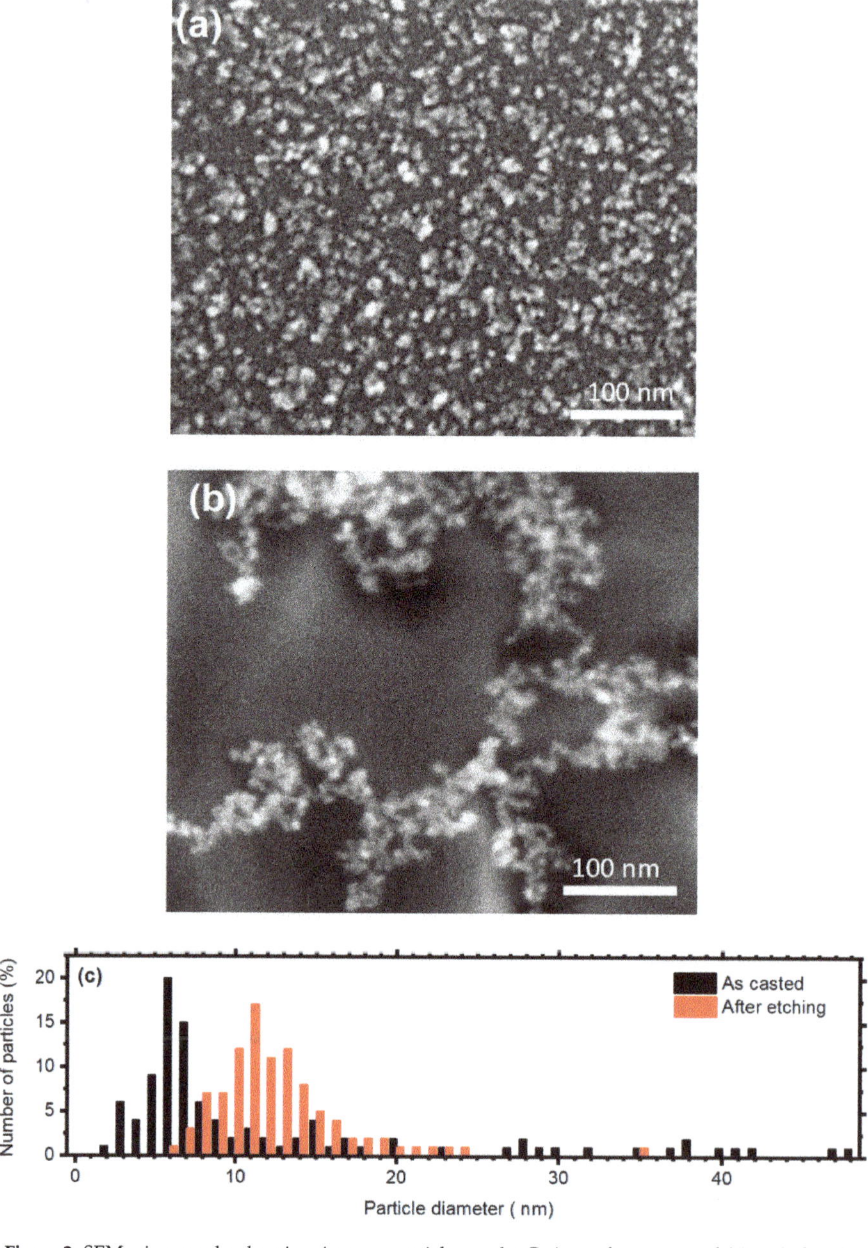

Figure 2. SEM micrographs showing Au nanoparticles on the GaAs surface as casted (**a**) and after the etching process. (**c**) Particles size distribution before (black) and after (**b**) GaAs MacEtch.

Figure 3 shows the SEM micrographs collected after etching of the (100) (panels a, a'), (110) (panels b, b'), (111)B (panels c, c'), and (211) (panels d, d') surfaces. Similar results were discussed in a previous work [12], but the brief analysis presented here provides a better interpretation of photon recycling mechanism of the etched surface. The top panels (a–d) of the figure display the cross-sectional micrographs collected for the different crystalline planes, while the bottom panels (a′–d′) show a tilted view at lower magnification. Etching of the (100) GaAs crystalline plane provides facets with ca. 25° characteristic angle with the sample plane and size of about 30 μm (Figure 3a). While these facets are partly exposed on the sample surface, the micrograph also shows the formation of holes beneath the surface. Figure 3a′ shows that these facets cover a large area, giving a homogeneous appearance to the surface. Figure 3b,b′ displays the results for the (110) plane. It is clear from Figure 3b that the surface exposes facets with similar orientation and size than those observed for the (100) surface. Panel b′ shows again a homogeneous distribution of etched planes on the entire sample surface. In the case of the (111)B surface, the etching generates randomly oriented hillocks (Figure 3c) and a homogenous surface (Figure 3c′). The facets on the etched (111)B orientation form larger angles than those observed in the (100) and (110) planes. Etching of the (211) plane digs conic-like holes into the surface and some spikes. These structures form again characteristic angles with respect to the sample plane (Figure 3d). Additionally, in this case, the resulting microstructures are homogeneously distributed and appear similar to those formed on the (111)B surface (Figure 3d′). Comparing the features size, it is possible to notice that the (111) surface provides less deep features than those arising from (211), (100), and (110) ones, which are instead similar in size. We suggest that the three surfaces showing deeper features are indeed etched at a faster rate, as monoatomic low-rate gallium planes are exposed to the etchant only at a later stage (i.e., the visible features in Figure 3a,b,d). Conversely, the etching of the highly reactive (111)B plane exposes the underlying (111)A plane (see Figure 1) where gallium atoms with mild reactivity can slow down the process.

The optical characteristics of the samples can be understood in terms of geometrical light trapping. The feature size, of the order of a few micrometers, is too large relative to visible and near infrared wavelengths to be treated within the effective medium theory. Light is therefore trapped within the etched surfaces upon multiple specular reflections. Figure 3e compares the reflectance spectra of the four etched samples with the one of a polished GaAs surface. The reflectance of polished GaAs approaches 0.7 in the range between 1100 and 925 nm. Moving toward shorter wavelengths, the values decreases to about 0.6 below 925 nm and remains rather constant until 700 nm where it increases and reaches ~0.85 at 550 nm. Conversely, the reflectance of etched GaAs approaches zero in the entire spectral range. The spectra are better visible in Figure 3f, where the intensity scale has been expanded. Within the GaAs transparency region, the minimum reflectance values are 0.013, 0.01, 0.008, and 0.007 for the (100), (110), (111)B, and (211) surfaces, respectively. Therefore, in the best case, MacEtch induces a 120-fold reduction of light reflection for the (211) surface at 550 nm (0.85/0.007).

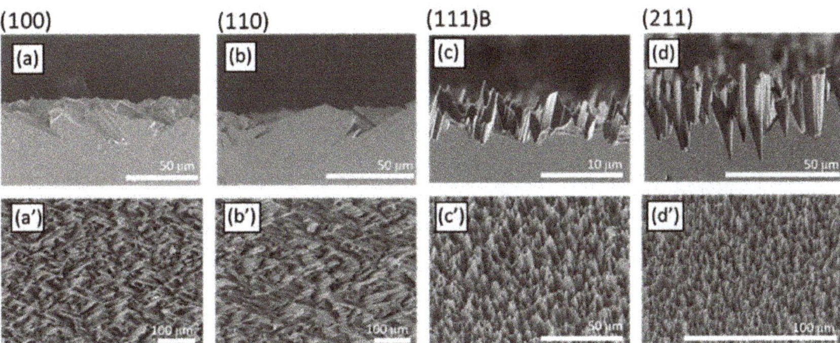

Figure 3. *Cont.*

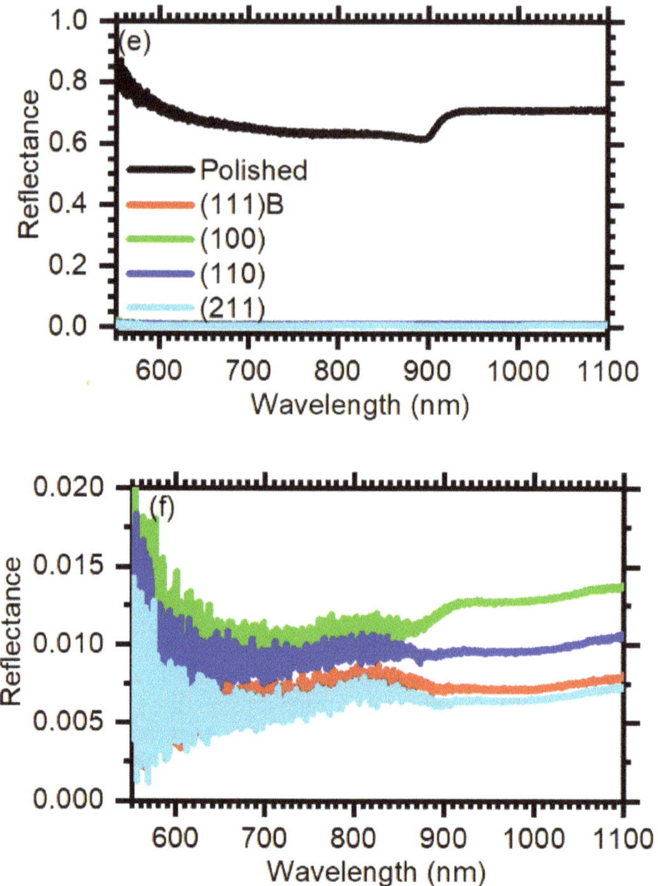

Figure 3. Cross-sectional (**a–d**) and tilted view (**a'–d'**) SEM micrographs of the (100) (**a,a'**), (110) (**b,b'**), (111)B (**c,c'**), and (211) (**d,d'**) GaAs crystalline planes. (**e**) Normal incidence reflectance spectra of polished (black) and etched GaAs surfaces: (100), red; (110), green; (111)B, blue; (211), cyan. (**f**) Reflectance spectra of the etched samples: A magnification of the reflectance scale allows to appreciate the spectral features.

Because the presence of residual gold would be detrimental for devices, we tested the possibility to dissolve the particles through a second etching step in aqua regia. Figure 4a reports the cross-sectional micrograph of the (111)B surface after the two MacEtch processes (see experimental section). There, gold is barely visible in the sample. However, several features on the surface appears broken. To reduce this detrimental effect, we also tested diluted etchants. Figure 4b shows the micrograph of a sample after etching with the acidic solution diluted 1:5 with water. Also in this case gold is barely visible, while the damages to the features are less evident than in the previous case. After further diluting the solution (1:10), the features are not damaged at all, but gold nanoparticles are still clearly visible in the micrograph as bright spots at the base of the etched features (Figure 4c). Then, we can infer that the dilution 1:5 can successfully remove the gold particles at a low expense for the microstructure quality.

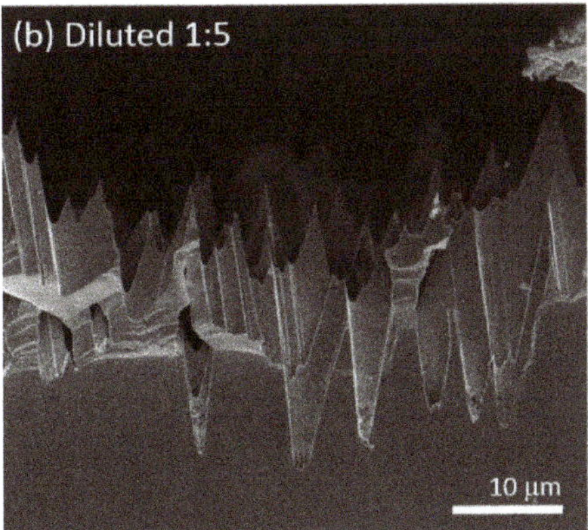

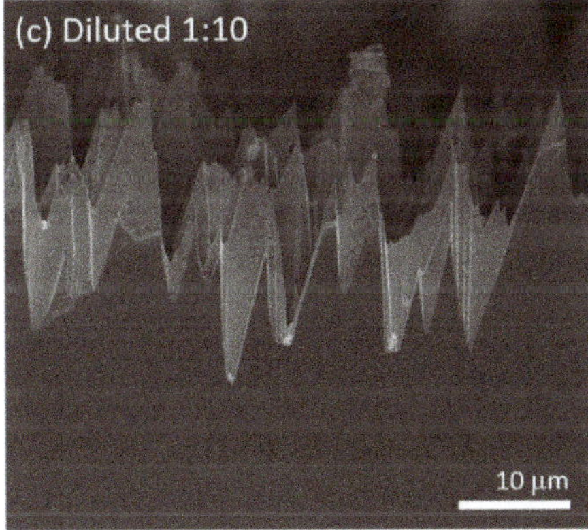

Figure 4. Cross-sectional SEM micrographs of the (111)B etched surface after 30 min treatment in pure aqua regia (**a**), and in aqua regia diluted in water with proportion 1:5 (**b**) and 1:10 (**c**).

As mentioned earlier, light trapping is a promising mechanism for photon recycling in hybrid polymer-inorganic optoelectronic devices. To demonstrate that the etched GaAs favors reabsorption of emitted photons, as a proof of principle, we cast a thin film of P3HT on the etched (111)B surface and investigated the resulting optical properties of the hybrid system. Figure 5a compares the reflectance spectra of the bare *black GaAs* (black line) with the one collected for *black GaAs* covered with P3HT (red line). The presence of P3HT increases the overall reflectance. While the bare *black GaAs* reflectance approaches zero in the entire spectral range, addition of the P3HT film generates a structured spectrum. The data show a maximum at 1100 nm, where the reflectance approaches 0.15. Moving toward shorter wavelengths, the intensity decreases slowly until ca. 870 nm. At energies above the GaAs energy gap, the reflectance then decreases with a faster rate until 660 nm, where it approaches zero. On the short wavelength side, a broad peak is detected between 420 and 660 nm, with a maximum at 505 nm and structures at 470, 570, and 611 nm, which are assigned to P3HT [6,63–65]. Overall, these data confirm the effective coverage of the surface by the polymer. To assess the effectiveness of photon recycling, we excited both P3HT and GaAs at 500 nm and compared the emission spectra of flat and etched surfaces. Photon recycling is expected to yield reduction of the P3HT photoluminescence and enhancement of the GaAs photoluminescence signals upon photon reabsorption in GaAs, favored by the trapping geometry. Figure 5b compares the emission spectrum collected for bare P3HT (green line) with P3HT cast on polished (red line) and *black GaAs* (black line). The bare P3HT emission spectrum ranges between 600 nm and 900 nm with a maximum of intensity at ~720 nm (green line) [66,67]. In the polished GaAs/P3HT spectrum no other features than those assigned to P3HT are evident and the emission intensity is lowered, suggesting electron transfer at the P3HT:GaAs interface (red line) [6]. The *black GaAs*/P3HT sample is characterized by stronger P3HT photoluminescence reduction compared to the flat sample, and by the presence of a peak at ~875 nm assigned to GaAs emission (black line) [12,33,68,69]. Quenching of the P3HT emission in the etched sample is consistent with geometrical light trapping discussed previously. Indeed, as sketched in Figure 5c, photons emitted from P3HT undergo several reflections between the surface features before they can leave the system, favoring reabsorption in GaAs. Because P3HT emission occurs below the HOMO-LUMO transition of the polymer, emitted photons cannot be reabsorbed by P3HT and therefore excite the underlying GaAs.

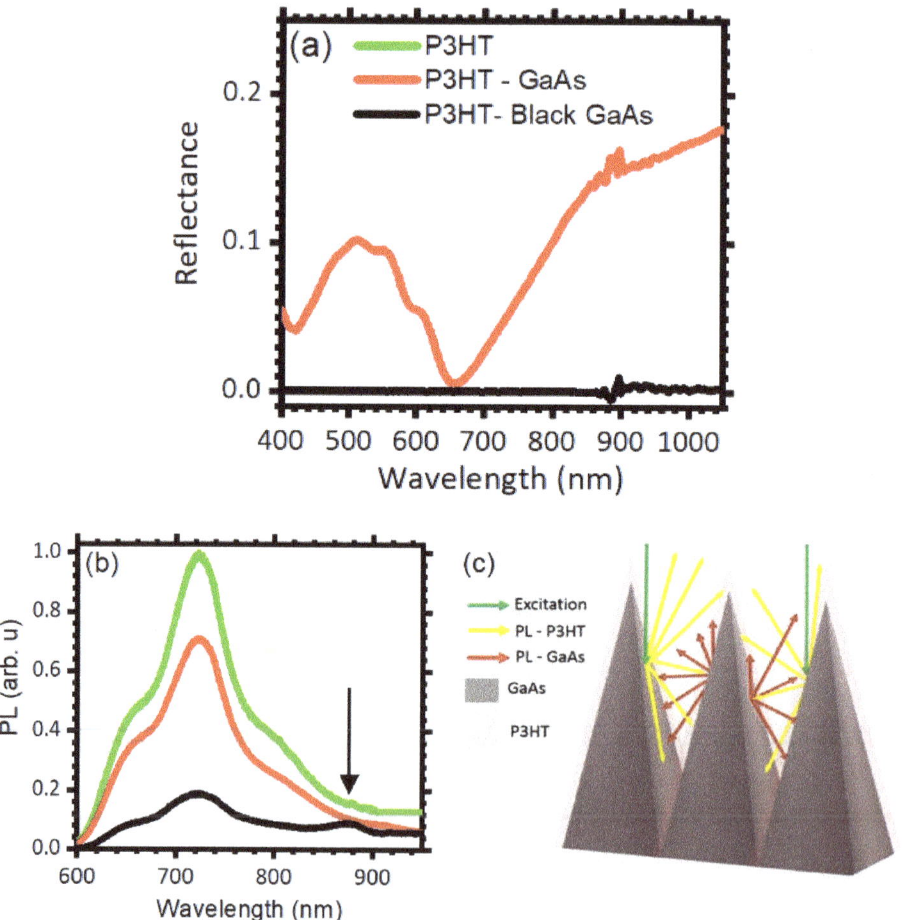

Figure 5. (**a**) Reflectance spectra collected for polished (red) and etched (black) GaAs (111)B surfaces covered with a P3HT thin layer. (**b**) Photoluminescence spectra collected for a P3HT film cast on a quartz substrate (green line), on polished GaAs (red line), and on *black GaAs* (black line). (**c**) Scheme of photon recycling at the *black GaAs*–P3HT interfaces.

These results are an important demonstration of a possible application of *black GaAs* as a photoactive material in hybrid solar cells. The superior semiconducting properties of GaAs, together with efficient light trapping and reabsorption of emitted photons shown here, could be employed to devise various hybrid structures for high performance photovoltaic devices.

4. Conclusions

We demonstrated that microstructured GaAs surfaces provide an efficient photon recycling platform to reduce losses associated with light emission occurring upon charge recombination in hybrid polymer-inorganic heterointerfaces. The microstructures were obtained by wet chemical etching of different GaAs crystalline planes in HF/H_2O_2, catalyzed by gold nanoparticles dispersed randomly on the semiconductor by electrodeless deposition. The process favors suppression of light reflection from the GaAs surface, with up to a 120-fold reduction. These are the best antireflective properties achieved for wet-etched semiconductors. Investigation of photoluminescence properties of GaAs:P3HT heterointerfaces confirmed that the microstructured GaAs reabsorb photons emitted by the P3HT cast on its surface. On the whole, the MacEtch process could be added to the fabrication workflow of photodetectors and solar cell devices to further improve their characteristics.

Author Contributions: Conceptualization, C.S.; methodology, P.L.; data analysis, P.L.; writing—original draft preparation, P.L.; writing—review and editing, P.L. and C.S. All authors have read and agreed to the published version of the manuscript.

Funding: C.S. would like to acknowledge the support of NTU Research Grant No. M4082409.

Acknowledgments: We thank Davide Comoretto for his unconditional support and for the useful discussions regarding this work. We also thank Valentina Robbiano and Franco Cacialli for their help in characterizing other properties of these samples.

Conflicts of Interest: The authors declare no conflict of interest.

References

1. Savin, H.; Repo, P.; Von Gastrow, G.; Ortega, P.; Calle, E.; Garín, M.; Alcubilla, R. Black silicon solar cells with interdigitated back-contacts achieve 22.1% efficiency. *Nat. Nanotechnol.* **2015**, *10*, 624–628. [CrossRef]
2. Füchsel, K.; Kroll, M.; Otto, M.; Steglich, M.; Bingel, A.; Käsebier, T.; Wehrspohn, R.B.; Kley, E.-B.; Pertsch, T.; Tünnermann, A. Black Silicon Photovoltaics. *Photon. Manag. Sol. Cells* **2015**, *3*, 117–151. [CrossRef]
3. Li, Z.; Guan, X.; Hua, Y.; Hui, S. Composite nanostructures induced by water-confined femtosecond laser pulses irradiation on GaAs/Ge solar cell surface for anti-reflection. *Opt. Laser Technol.* **2018**, *106*, 222–227. [CrossRef]
4. Tang, Z.; Tress, W.R.; Inganäs, O. Light trapping in thin film organic solar cells. *Mater. Today* **2014**, *17*, 389–396. [CrossRef]
5. Chang, K.W.; Sun, K.W. Highly efficient back-junction PEDOT: PSS/n-Si hybrid solar cell with omnidirectional antireflection structures. *Org. Electron.* **2018**, *55*, 82–89. [CrossRef]
6. Yin, J.; Migas, D.B.; Panahandeh-Fard, M.; Chen, S.; Wang, Z.; Lova, P.; Soci, C. Charge redistribution at GaAs/P3HT heterointerfaces with different surface polarity. *J. Phys. Chem. Lett.* **2013**, *4*, 3303–3309. [CrossRef]
7. Zhang, W.; Anaya, M.; Lozano, G.; Calvo, M.E.; Johnston, M.B.; Míguez, H.; Snaith, H.J. highly efficient Perovskite solar cells with tunable structural color. *Nano Lett.* **2015**, *15*, 1698–1702. [CrossRef]
8. Iasilli, G.; Francischello, R.; Lova, P.; Silvano, S.; Surace, A.; Pesce, G.; Alloisio, M.; Patrini, M.; Shimizu, M.; Comoretto, D.; et al. Luminescent solar concentrators: Boosted optical efficiency by polymer dielectric mirrors. *Mater. Chem. Front.* **2019**, *3*, 429–436. [CrossRef]
9. Garnett, E.C.; Yang, P. Light Trapping in Silicon Nanowire Solar Cells. *Nano Lett.* **2010**, *10*, 1082–1087. [CrossRef]
10. Da, Y.; Xuan, Y.; Li, Q. From light trapping to solar energy utilization: A novel photovoltaic–thermoelectric hybrid system to fully utilize solar spectrum. *Energy* **2016**, *95*, 200–210. [CrossRef]
11. Brongersma, M.L.; Cui, Y.; Fan, S. Light management for photovoltaics using high-index nanostructures. *Nat. Mater.* **2014**, *13*, 451–460. [CrossRef] [PubMed]
12. Lova, P.; Robbiano, V.; Cacialli, F.; Comoretto, D.; Soci, C. Black GaAs by metal-assisted chemical etching. *ACS Appl. Mater. Interfaces* **2018**, *10*, 33434–33440. [CrossRef] [PubMed]

13. Lova, P.; Manfredi, G.; Comoretto, D. Advances in functional solution processed planar one-dimensional photonic crystals. *Adv. Opt. Mater.* **2018**, *6*, 1800730. [CrossRef]

14. Lova, P.; Soci, C. Nanoimprint lithography: Toward polymer photonic crystals. In *Organic and Hybrid Photonic Crystals*, 1st ed.; Comoretto, D., Ed.; Springer International Publishing: Cham, Switzerland, 2015; Volume 1, p. 493.

15. Lova, P. Selective polymer distributed bragg reflector vapor sensors. *Polymers* **2018**, *10*, 1161. [CrossRef] [PubMed]

16. Miller, O.D.; Yablonovitch, E.; Kurtz, S.R. Strong internal and external luminescence as solar cells approach the Shockley–Queisser limit. *IEEE J. Photovolt.* **2012**, *2*, 303–311. [CrossRef]

17. Li, X.; Bohn, P.W. Metal-assisted chemical etching in HF/H$_2$O$_2$ produces porous silicon. *Appl. Phys. Lett.* **2000**, *77*, 2572–2574. [CrossRef]

18. Chartier, C.; Bastide, S.; Levy-Clement, C. Metal-assisted chemical etching of silicon in HF–H2O2. *Electrochim. Acta* **2008**, *53*, 5509–5516. [CrossRef]

19. Peng, K.-Q.; Lu, A.; Zhang, R.; Lee, S.-T. Motility of metal nanoparticles in silicon and induced anisotropic silicon etching. *Adv. Funct. Mater.* **2008**, *18*, 3026–3035. [CrossRef]

20. Zhang, M.; Peng, K.-Q.; Fan, X.; Jie, J.; Zhang, R.-Q.; Lee, S.-T.; Wong, N.-B. Preparation of large-area uniform silicon nanowires arrays through metal-assisted chemical etching. *J. Phys. Chem. C* **2008**, *112*, 4444–4450. [CrossRef]

21. Jansen, H.; De Boer, M.J.; Unnikrishnan, S.; Louwerse, M.C.; Elwenspoek, M.C. Black silicon method: X. A review on high speed and selective plasma etching of silicon with profile control: An in-depth comparison between Bosch and cryostat DRIE processes as a roadmap to next generation equipment. *J. Micromech. Microeng.* **2009**, *19*, 33001. [CrossRef]

22. Srivastava, S.; Kumar, D.; Singh, P.; Kar, M.; Kumar, V.; Husain, M. Excellent antireflection properties of vertical silicon nanowire arrays. *Sol. Energy Mater. Sol. Cells* **2010**, *94*, 1506–1511. [CrossRef]

23. Huang, Z.; Geyer, N.; Werner, P.; De Boor, J.; Goesele, U. Metal-assisted chemical etching of silicon: A review. *Adv. Mater.* **2011**, *42*, 285–308. [CrossRef]

24. Agnieszka, K.; Seán, T.B. Metal-assisted chemical etching using sputtered gold: A simple route to black silicon. *Sci. Technol. Adv. Mater.* **2011**, *12*, 045001.

25. Smith, Z.R.; Smith, R.; Collins, S. Mechanism of nanowire formation in metal assisted chemical etching. *Electrochim. Acta* **2013**, *92*, 139–147. [CrossRef]

26. Beetz, B. Suntech's Black Silicon Solar Cells Enter Mass Production. *PV Magazine.* 4 January 2018. Available online: https://www.pv-magazine.com/2018/01/04/suntechs-black-silicon-solar-cells-enter-mass-production/ (accessed on 4 June 2020).

27. Black Silicon Solar Modules: A Powerful and Pleasing Design. *Trina Solar.* 17 November 2015. Available online: https://www.trinasolar.com/us/resources/blog/black-silicon-solar-modules-powerful-and-pleasing-design (accessed on 27 May 2020).

28. Rezvani, S.J.; Pinto, N.; Boarino, L. Rapid formation of single crystalline Ge nanowires by anodic metal assisted etching. *Cryst. Eng. Comm.* **2016**, *18*, 7843–7848. [CrossRef]

29. Elwenspoek, M.; Lindberg, U.; Kok, H.; Smith, L. Wet chemical etching mechanism of silicon. In Proceedings of the IEEE Workshop on Micro Electro Mechanical Systems, MEMS 1994, Oiso, Japan, 25–28 January 1994; pp. 223–228.

30. Baca, A.G.; Ashby, C.I.H. *Fabrication of GaAs Devices*; Institution of Engineering and Technology: Stevenage, UK, 2005; p. 370.

31. Unagami, T. Formation mechanism of porous silicon layer by anodization in HF solution. *J. Electrochem. Soc.* **1980**, *127*, 476. [CrossRef]

32. DeJarld, M.; Shin, J.C.; Chern, W.; Chanda, D.; Balasundaram, K.; Rogers, J.A.; Li, X. formation of high aspect ratio GaAs nanostructures with metal-assisted chemical etching. *Nano Lett.* **2011**, *11*, 5259–5263. [CrossRef]

33. Mohseni, P.K.; Kim, S.H.; Zhao, X.; Balasundaram, K.; Kim, J.D.; Pan, L.; Rogers, J.A.; Coleman, J.; Li, X. GaAs pillar array-based light emitting diodes fabricated by metal-assisted chemical etching. *J. Appl. Phys.* **2013**, *114*, 64909. [CrossRef]

34. Asoh, H.; Imai, R.; Hashimoto, H. Au-Capped GaAs Nanopillar arrays fabricated by metal-assisted chemical etching. *Nanoscale Res. Lett.* **2017**, *12*, 444. [CrossRef]

35. Asoh, H.; Suzuki, Y.; Ono, S. Metal-assisted chemical etching of GaAs using Au catalyst deposited on the backside of a substrate. *Electrochim. Acta* **2015**, *183*, 8–14. [CrossRef]

36. Lee, A.R.; Kim, J.; Choi, S.-H.; Shin, J.C. Formation of three-dimensional GaAs microstructures by combination of wet and metal-assisted chemical etching. *Phys. Statu. Solidi RRL* **2014**, *8*, 345–348. [CrossRef]

37. Ono, S.; Kotaka, S.; Asoh, H. Fabrication and structure modulation of high-aspect-ratio porous GaAs through anisotropic chemical etching, anodic etching, and anodic oxidation. *Electrochim. Acta* **2013**, *110*, 393–401. [CrossRef]

38. Song, Y.; Oh, J. Fabrication of three-dimensional GaAs antireflective structures by metal-assisted chemical etching. *Sol. Energy Mater. Sol. Cells* **2016**, *144*, 159–164. [CrossRef]

39. Cowley, A.; Steele, J.A.; Byrne, D.; Vijayaraghavan, R.K.; McNally, P.J. Fabrication and characterisation of GaAs nanopillars using nanosphere lithography and metal assisted chemical etching. *RSC Adv.* **2016**, *6*, 30468–30473. [CrossRef]

40. Kong, L.; Song, Y.; Kim, J.D.; Yu, L.; Wasserman, D.; Chim, W.K.; Chiam, S.Y.; Li, X. Damage-free smooth-sidewall in GaAs nanopillar array by metal-assisted chemical etching. *ACS Nano* **2017**, *11*, 10193–10205. [CrossRef]

41. Wilhelm, T.S.; Kolberg, A.P.; Baboli, M.A.; Abrand, A.; Bertness, K.A.; Mohseni, P.K. Communication—black GaAs with sub-wavelength nanostructures fabricated via lithography-free metal-assisted chemical etching. *ECS J. Solid State Sci. Technol.* **2019**, *8*, Q134–Q136. [CrossRef]

42. Wilhelm, T.S.; Wang, Z.; Baboli, M.A.; Yan, J.; Preble, S.F.; Mohseni, P.K. Ordered AlxGa1−xAs nanopillar arrays via inverse metal-assisted chemical etching. *ACS Appl. Mater. Interfaces* **2018**, *10*, 27488–27497. [CrossRef]

43. Wilhelm, T.S.; Soule, C.W.; Baboli, M.A.; O'Connell, C.J.; Mohseni, P.K. Fabrication of Suspended III–V nanofoils by inverse metal-assisted chemical etching of in 0.49 Ga 0.51P/GaAs heteroepitaxial films. *ACS Appl. Mater. Interfaces* **2018**, *10*, 2058–2066. [CrossRef]

44. Green, M.A.; Views, C. Solar Cell Efficiency Tables. *Chem. Views* **2012**, *27*, 565–575. [CrossRef]

45. Blakemore, J.S. Semiconducting and other major properties of gallium arsenide. *J. Appl. Phys.* **1982**, *53*, R123–R181. [CrossRef]

46. Song, Y.; Choi, K.; Jun, D.-H.; Oh, J. Nanostructured GaAs solar cells via metal-assisted chemical etching of emitter layers. *Opt. Express* **2017**, *25*, 23862. [CrossRef] [PubMed]

47. Glinsner, T.; Kreindl, G. *Lithography*; Wang, M., Ed.; Intech Open: London, UK, 2010; p. 656. Available online: https://www.intechopen.com/books/lithography (accessed on 4 June 2020).

48. Peng, K.-Q.; Zhang, M.; Lu, A.; Wong, N.-B.; Zhang, R.; Lee, S.-T. Ordered silicon nanowire arrays via nanosphere lithography and metal-induced etching. *Appl. Phys. Lett.* **2007**, *90*, 163123. [CrossRef]

49. Wilhelm, T.S.; Kolberg, A.P.; Wang, Z.; Soule, C.W.; Baboli, M.A.; Yan, J.; Preble, S.F.; Mohseni, P.K. Metal-assisted chemical etching for simple, cost-effective, and large-scale III-V semiconductor nanofabrication. In Proceedings of the 235th ECS Meeting, Dallas, TX, USA, 26–30 May 2019.

50. Peng, K.-Q.; Yan, Y.; Gao, S.-P.; Zhu, J. Dendrite-assisted growth of silicon nanowires in electroless metal deposition. *Adv. Funct. Mater.* **2003**, *13*, 127–132. [CrossRef]

51. Geyer, N.; Fuhrmann, B.; Leipner, H.; Werner, P. Ag-mediated charge transport during metal-assisted chemical etching of silicon nanowires. *ACS Appl. Mater. Interfaces* **2013**, *5*, 4302–4308. [CrossRef] [PubMed]

52. MacFadyen, D.N. On the Preferential Etching of GaAs by H_2SO_4-H_2O_2-H_2O J. *Electrochem. Soc.* **1983**, *130*, 1934–1941. [CrossRef]

53. Bienaime, A.; Elie-Caille, C.; Leblois, T. Micro structuration of gas surface by wet etching: Towards a specific surface behavior. *J. Nanosci. Nanotechnol.* **2012**, *12*, 6855–6863. [CrossRef]

54. Yasukawa, Y.; Asoh, H.; Ono, S. Morphological control if periodic GaAs hole arrays by simple au-mediated wet etching. *J. Electrochem. Soc.* **2012**, *159*, CD328–CD332. [CrossRef]

55. Li, X. Metal assisted chemical etching for high aspect ratio nanostructures: A review of characteristics and applications in photovoltaics. *Curr. Opin. Solid State Mater. Sci.* **2012**, *16*, 71–81. [CrossRef]

56. Cheung, H.-Y.; Lin, H.; Xiu, F.; Wang, F.; Yip, S.; Ho, J.C.; Wong, C.-Y. Mechanistic characteristics of metal-assisted chemical etching in GaAs. *J. Phys. Chem. C* **2014**, *118*, 6903–6908. [CrossRef]

57. Song, Y.; Oh, J. Thermally driven metal-assisted chemical etching of GaAs with in-position and out-of-position catalyst. *J. Mater. Chem. A* **2014**, *2*, 20481–20485. [CrossRef]

58. Li, G.; Zhu, R.; Yang, Y. Polymer solar cells. *Nat. Photon.* **2012**, *6*, 153–161. [CrossRef]

59. Sheng, X.; Yun, M.H.; Zhang, C.; Al-Okaily, A.M.; Masouraki, M.; Shen, L.; Wang, S.; Wilson, W.L.; Kim, J.Y.; Ferreira, P.; et al. Solar cells: Device architectures for enhanced photon recycling in thin-film multijunction solar cells. *Adv. Energy Mater* **2015**, *5*, 1400919. [CrossRef]

60. Saliba, M.; Zhang, W.; Burlakov, V.M.; Stranks, S.D.; Sun, Y.; Ball, J.M.; Johnston, M.B.; Goriely, A.; Wiesner, U.; Snaith, H.J. Plasmonic-induced photon recycling in metal halide perovskite solar cells. *Adv. Funct. Mater.* **2015**, *25*, 5038–5046. [CrossRef]

61. Momma, K.; Izumi, F. VESTA 3 for three-dimensional visualization of crystal, volumetric and morphology data. *J. Appl. Crystallogr.* **2011**, *44*, 1272–1276. [CrossRef]

62. Crystallography Open Database. Available online: http://www.crystallography.net/result.php (accessed on 21 May 2018).

63. Yang, C.; Soci, C.; Moses, D.; Heeger, A. Aligned rrP3HT film: Structural order and transport properties. *Synth. Met.* **2005**, *155*, 639–642. [CrossRef]

64. Bi, H.; Lapierre, R.R. A GaAs nanowire/P3HT hybrid photovoltaic device. *Nanotechnology* **2009**, *20*, 465205. [CrossRef]

65. Xie, X.; Ju, H.; Lee, E.-C.; Xiaoyin, X.; Heongkyu, J.; Eun-Cheol, L. Band gap enhancement by covalent interactions in P3HT/PCBM photovoltaic heterojunction. *J. Korean Phys. Soc.* **2010**, *57*, 144–148. [CrossRef]

66. Kim, Y.; Bradley, D.D.C. Bright red emission from single layer polymer light-emitting devices based on blends of regioregular P3HT and F8BT. *Curr. Appl. Phys.* **2005**, *5*, 222–226. [CrossRef]

67. Brown, P.J.; Thomas, S.; Köhler, A.; Wilson, J.S.; Kim, J.-S.; Ramsdale, C.M.; Sirringhaus, H.; Friend, R.H. Effect of interchain interactions on the absorption and emission of poly(3-hexylthiophene). *Phys. Rev. B* **2003**, emph67, 064203. [CrossRef]

68. Takao, N.; Kyo-ichiro, F.; Taiji, O. Improvement of Quantum Efficiency in Gallium Arsenide Electroluminescent Diodes. *Jpn. J. Appl. Phys.* **1967**, *6*, 665.

69. Sabataityt, J.; Šimkien, I.; Bendorius, R.-A.; Grigoras, K.; Jasutis, V.; Pačebutas, V.; Tvardauskas, H.; Naudzius, K. Morphology and strongly enhanced photoluminescence of porous GaAs layers made by anodic etching. *Mater. Sci. Eng. C* **2002**, *19*, 155–159. [CrossRef]

© 2020 by the authors. Licensee MDPI, Basel, Switzerland. This article is an open access article distributed under the terms and conditions of the Creative Commons Attribution (CC BY) license (http://creativecommons.org/licenses/by/4.0/).

 *micromachines*

Article

Silicon Conical Structures by Metal Assisted Chemical Etching

Oscar Pérez-Díaz and Enrique Quiroga-González *

Institute of Physics, Benemérita Universidad Autónoma de Puebla, Puebla 72570, Mexico; oscpedi@gmail.com
* Correspondence: equiroga@ieee.org; Tel.: +52-222-229-5610

Received: 27 March 2020; Accepted: 10 April 2020; Published: 11 April 2020

Abstract: A simple and inexpensive method to obtain Si conical structures is proposed. The method consists of a sequence of steps that include photolithography and metal assisted chemical etching (MACE) to create porous regions that are dissolved in a post-etching process. The proposed process takes advantage of the lateral etching obtained when using catalyst particles smaller than 40 nm for MACE. The final shape of the base of the structures is mainly given by the shape of the lithography mask used for the process. Conical structures ranging from units to hundreds of microns can be produced by this method. The advantage of the method is its simplicity, allowing the production of the structures in a basic chemical lab.

Keywords: silicon cones; metal assisted chemical etching; transversal pores

1. Introduction

Different techniques have been developed in order to produce Si structures to be used in applications such optoelectronics [1], energy storage [2], or sensors [3]. Among these techniques it is possible to find reactive ion etching (RIE), inductively-coupled plasma (ICP)-RIE, or chemical assisted ion beam etching (CAIBE); however, all of them require special equipment, like vacuum chambers or plasma generators. On the other hand, electrochemical etching of Si has proved to be a good option because it provides high control for Si dissolution. Nevertheless, etching complete wafers is complex, since high current densities in the range of amperes are required, which derive in undesired heating that makes necessary the use of high-quality cooling appliances.

On the other hand, the metal assisted chemical etching (MACE) technique does not require any special equipment or facilities, and makes possible the fabrication of complex structures [4–6]. MACE is performed by immersing a piece of semiconductor (commonly Si, but also other semiconductors like Ge [7] or III-V semiconductors [8,9] can be used), previously coated with a catalyst (usually the metals Au, Pt, or Ag [10]), in an HF based solution containing an oxidant agent (commonly H_2O_2, $Na_2S_2O_8$ or $KMnO_4$ [11,12]). The metal catalyzes the release of electronic holes from the oxidant and, depending on the potential energy difference between metal and semiconductor, it promotes their injection to the semiconductor. In the most common etching case, Si is oxidized beneath or around the metallized sections, and this oxide is dissolved by HF.

Despite it is possible to obtain different structures using MACE, the most of the reports on this technique indicate that the etching occurs most probably in crystallographic directions (there are fast etching planes and etch-stopper planes) or vertically [13]. However, it is also known that when the catalyst particles have diameters smaller than 40 nm, they produce pores either vertically or horizontally [14,15]. Furthermore, when MACE has been performed using particles with a dispersion of sizes from 10 to 400 nm, thin pores (with diameters below 100 nm) and wide pores (with diameters of hundreds of nanometers) were obtained vertically, while just thin pores were obtained horizontally. By eliminating the most of the particles of sizes below 100 nm, mainly vertical porosification was obtained [16]. The vertical porosification of large particles is due to the much larger contact area below than at the sides of these particles, considering that they are spherical. For lightly-doped p-type Si wafers, for example, a larger contact area means a larger injection of electronic holes, which speeds the etching rate up. The contact area plays an important role in MACE [17]. When the particles are smaller, the probability is the same to etch either vertically or horizontally. With this equal probability, it is possible to think about a porous section growing upon the time in a direction with an angle close to 45° with respect to the vertical (the same amount of particles may move in the x than in the y direction, at similar velocity, producing that dy/dx = 1). To the knowledge of the authors, there are no reports to date taking advantage of this effect.

In the present work, a methodology to obtain Si conical structures by MACE is proposed. It consist of creating porous regions by MACE (previously defined by photolithography), which grow in angle with the time (due to lateral and vertical porosification obtained with catalyst particles smaller than 40 nm), and removing them afterwards. This methodology brings more flexibility to the MACE process. Additionally, it has the advantage of being simple, because the entire process could be performed in basic chemical labs without the need of complex facilities or equipment. Arrays of conical structures are important for different applications. They have been used as multielectrode sensing platforms for neuronal or cardiac tissue [18]. Additionally, arrays of complete or truncated cones have been used as antireflection layers [19] or to enhance the absorption of light [20], for different optical and optoelectronic applications such as solar cells. Moreover, such arrays have been used to modify the wetting properties of surfaces [21], achieving even super-repellency of hydrophobic surfaces [22].

2. Materials and Methods

p-type (100) Si wafers with resistivity of 15–25 $\Omega \cdot$cm were used as starting material. The fabrication procedure to obtain the conical structures of the present work consists of a sequence of steps: (a) Photolithography, (b) chemical deposition of Ag particles, (c) MACE etching, (d) dry oxidation of the porosified sections, and (e) dissolution of oxide. Alternatively, at the end of the process one can also dissolve the Ag particles in solutions of HNO_3. The steps are schematized in Figure 1.

A quadratic pattern of circles was transferred by photolithography to a film of photoresist previously deposited on the Si wafers. The photoresist acts as masking layer for the metal deposition. The metallization can be done as sophisticated and controlled as in the case of thin film deposition by sputtering [23–25], or as simple as in the case of chemical deposition of metal particles using just a beaker [26]. For this report, it was used the simplest case. Ag particles were chemically deposited on the uncovered sections of Si by immersion in a solution 0.1 mM of $AgNO_3$ in a mixture of HF (48%), H_2O_2 (30%) and H_2O, in a proportion 2:3.4:94.6 *v/v*. High-density polypropylene beakers were used for this and all the subsequent processes with HF, since that material endures adequately this acid. The deposition time was 90 s, being performed in an ultrasonic bath in order to obtain a homogeneous distribution of particles.

The etching process was performed using an aqueous solution containing HF (48%), H_2O_2 (30%) as oxidant agent [27], and deionized water (DI), in a proportion 4:7:40 *v/v* at 30 °C. The etching time was 5 h. With this process, porous Si sections were obtained. In order to obtain the final structures, the porous sections of the Si samples need to be dissolved. To accomplish this, it is possible to use anisotropic [28,29] or isotropic chemical etching techniques [30]; however, in order to dissolve mainly the porous sections without important modification of the shape of the remaining Si, those techniques were avoided in this work. The samples were submitted to thermal oxidation at 850 °C under O_2 flux (1 sccm) for 3 h. With this process, the porous Si sections were oxidized. To dissolve the oxide, the samples were immersed in a solution of HF (48%) and H_2O in a proportion 1:9 *v/v* for 60 s. Silicon oxides are highly soluble in HF based solutions [31]. The final structures were analyzed with a JEOL JSM-7500F (Tokyo, Japan) field emission scanning electron microscope.

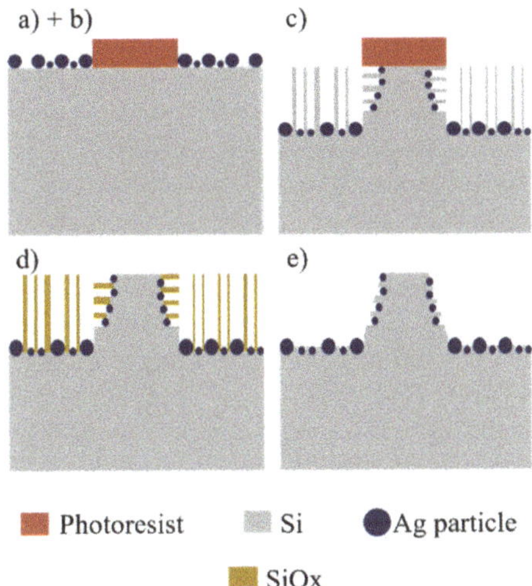

Figure 1. Schematic of the process to produce conical structures: (**a**) Photolithography, (**b**) deposition of Ag particles, (**c**) MACE, (**d**) dry oxidation, and (**e**) dissolution of oxide.

3. Results and Discussion

Figure 2a shows a SEM (scanning electron microscope) micrograph of a Si sample after Ag deposition. The Ag particles are the white sections in the micrograph. Their shape is semi-spherical, but sometimes the particles coalesce giving rise to ovoidal forms. Semi-spherical shapes are commonly obtained when depositing using low concentrated $AgNO_3$ solutions [32]; the particles nucleate and start to grow until they coalesce and could form dendrites at longer deposition times [16]. It is also important to note that the particles are encrusted in Si. This happens because of the use of H_2O_2 during the deposition process: The Ag particles deposit on Si and catalyze the etching of Si at the same time, in the presence of the oxidant. However, the trenches are shallow because the deposition time is short (90 s). The deposited Ag particles have diameters in the range of 10 to 70 nm. A histogram of the particle size distribution (measured from SEM micrographs of the deposits) is presented in Figure 2b. As can be seen, the most of the particles have sizes below 40 nm. It was intended to have particles with sizes below 40 nm taking into account previous studies that suggest that with particles of those sizes the probability of etching vertically or horizontally is similar [16,18,19].

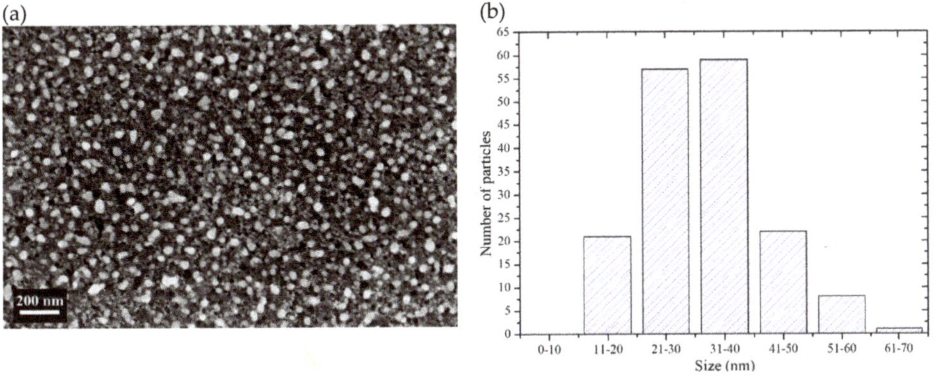

Figure 2. (**a**) SEM micrograph of the surface of a Si sample after deposition of Ag particles. (**b**) Size distribution of the Ag particles of the deposits.

Figure 3 shows a SEM micrograph of the structures obtained after the MACE process, dry oxidation and oxide dissolution. As can be observed, the structures are arrays of truncated cones. The bases and tops of the cones differ a bit from the circular shape. The diameters of the cones are 52 ± 5 μm for the top part and around 120 μm for the bottom part. They have a height of 60 μm.

Figure 4 shows a close-up to the structures. The walls of the cones is rough, with apparent porosity. This is an indication that the oxidation time was not enough to oxidize the whole porous Si sections. Because of this, the porous sections could not be completely dissolved during the last treatment in HF solutions (that dissolve SiO_2). However, the porosity of the cone walls is a good indication of the transversal porosification. Taking a look at the surface of the cones (inset of Figure 4) helps confirming the existence of transversal pores. They grow in the <100> directions. In principle, one would not expect pores exactly at the surface; however, Ag particles may grow through the photoresist (the photoresist is partially permeable to Ag^+ ions during the deposition of Ag particles). It is important to mention that the photoresist used for the experiments of the present work is not HF resistant; nevertheless, it stands enough time for the Ag deposition, and it starts detaching during the etching process. It is not necessary that the photoresist stands the whole etching time, since no masking layer is necessary for this process (the etching rate in sections with Ag is hundreds of times faster than the etching rate in sections without catalyst). The few Ag particles grown beneath the photoresist could move in the X-Y plane during the etching process due to the availability of etchant in the surface (the photoresist does not stand HF, and the acid could diffuse through or below it); for this reason, it is possible to see transversal pores exactly at the surface.

Figure 3. SEM micrograph of the obtained structures after the whole process. An array of truncated cones is evident.

Taking a closer look to one entire truncated cone (see Figure 5), one can observe two different slopes of the cone walls. Going up to down, the first slope is of 2.8, while the second is of 1.3. The steep first slope is given by an excess of etchant; thus, the etching process is reaction-rate limited. As could be observed in the histogram of Figure 2, there is a good number of particles larger than 40 nm. Those particles have a higher probability of etching vertically. As they offer larger areas to catalyze the decomposition of H_2O_2, they inject a larger number of electronic holes to the semiconductor enabling a faster etching rate than with the smaller particles. It is known that the one dimensional (vertical) etching rate increases with the catalyst particle size (in particular with the coverage area of the catalyst) [10], but until certain limit of sizes, when the mass transfer beneath the catalyst particles is limited, and the etching rate starts to decrease [11]. After the first 24 μm of etching in depth, the process is diffusion limited. It is common to observe diffusion limitation during a MACE process [33]. In this way, the etching process is mainly controlled by the availability of etchant, and the effect of the particle size is secondary. The difference between the vertical and the horizontal etching rate is about 30% in this depth range (27.5 µm of lateral etching vs. 36 µm of vertical etching, producing a slope of 1.3). The difference of etching rates in shallower depths is 180%. Following the tendency of the etching fronts, evoluting in angle, one can predict that if the MACE etching time is longer, complete conical structures (not truncated) could be obtained.

Figure 4. SEM micrograph closing up at the wall of the cones. Inset: Top view of the cones.

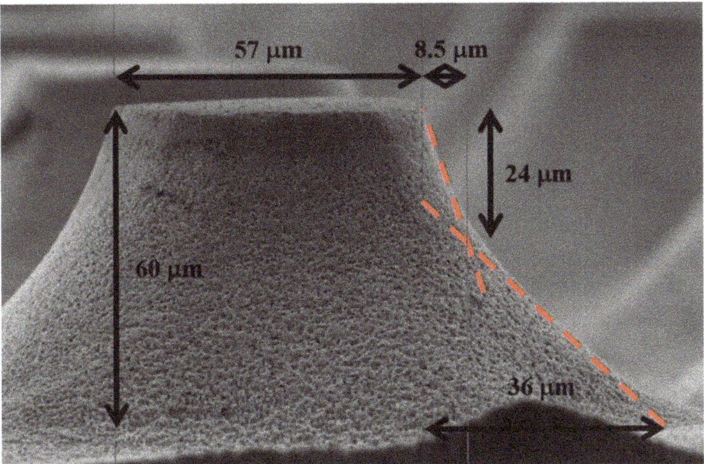

Figure 5. SEM micrograph of a truncated cone indicating its dimensions. The dashed lines indicate the two slopes of the cone walls.

Figure 6 shows a top view of the pattern of photoresist used during the etching process (photograph of the left), in contrast to the pattern of truncated cones obtained (SEM micrograph of the right). The photograph was captured with a portable optical microscope equipped with a CCD (charge-coupled device) camera. The dots of photoresist deviate a bit from the circular shape due to the resolution of the photomask, which was fabricated with a conventional paper printer. The diameter of the dots of the original pattern is of about 120 µm, with a pitch of 230 µm. The final structures have an upper diameter of 52 ± 5 µm, with a lower diameter in the range of 120 µm (as the original pattern).

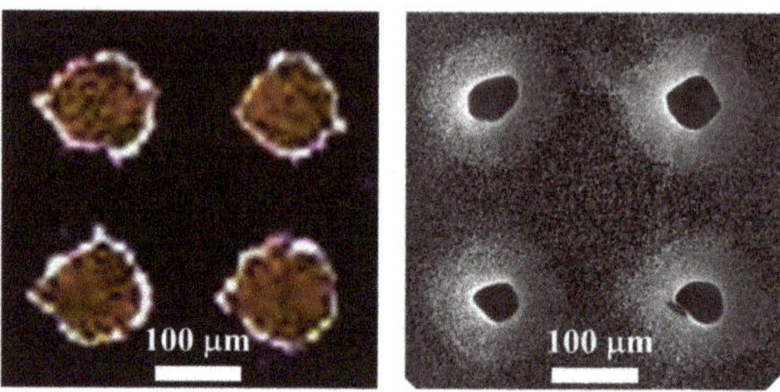

Figure 6. Left: Photograph of the pattern of photoresist used during the process. Right: SEM micrograph of a top view of the array of truncated cones.

The fact that the top of the cones does not have the same shape than the dots of photoresist could be explained with the fact that the etching occurs mainly in <100> directions (see Figure 4). The etching profile of lateral pores, saw from above, is schematized in Figure 7. The dots of the figure represent Ag particles, while the straight lines represent the pores. All the straight lines have the same length, considering equal etching rates in the [100] and [010] directions. It is clear that the region without lines (pores) is not exactly a circle. It is also possible to observe that there are sections with lines (pores) just in one direction, thus the density of pores in those sections is just the half. Sections with pores in two directions can be oxidized faster, due to the higher density of pores, which provide larger surface areas to be oxidized. These oxidized sections can be easily removed in HF solutions. One could still observe pores in the walls of the cones of this work. This should indicate that the oxidation of all porous sections was not complete (the oxidation of the sections with pores in just one direction takes longer). If the oxidation time would be reduced even more, the cones would tend to have a flower-like shape (the sum of the sections with no lines and the sections with lines in one direction of Figure 7).

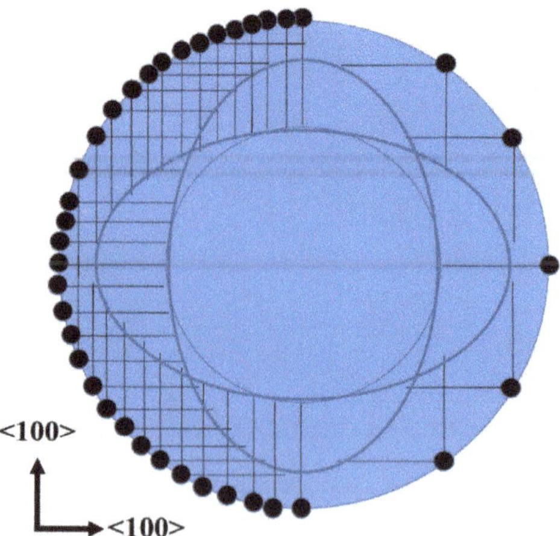

Figure 7. Schematic of the lateral pores observed from above. The black dots represent Ag particles, while the straight lines represent pores.

To prove that the proposed process to produce conical structures works also in the micron range, an experiment was performed using a mask with a quadratic array of circular dots of 1.5 µm, with pitch of 3 µm. A micrograph of the resulting structures is shown in Figure 8a. Figure 8b shows a micrograph of the cross-section of one of the micro-cones. As can be observed, no pores cross the structure, supporting the theory of cone formation. Despite the surface of the cones looks porous, the bulk is solid. The lines observed in the cross section are a cleavage artifact.

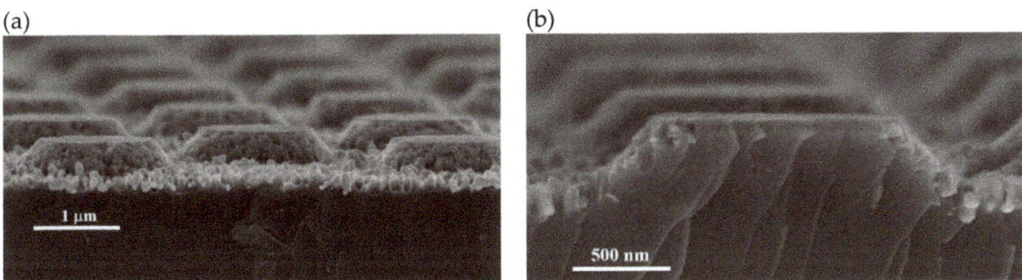

Figure 8. SEM micrographs of micron sized truncated cones produced by the methodology of this work. (**a**) Overview; (**b**) cross-section of a cone.

4. Conclusions

Transversal porosification of Si by MACE using Ag particles of sizes smaller of 40 nm has been used as basis to produce conical structures. At etching depths smaller than 24 µm, porosification is controlled by the reaction rate, producing steeper cone walls. Deeper etching is limited by the diffusion of the etchant, producing a reduction of the slope of the cone walls. Transversal porosification occurs mainly in the <100> direction. Due to this, the final cross-sectional shapes of the cones do not follow exactly the shape of the patterns of the photolithography mask. It was proved that the methodology works to produce conical structures of sizes from units to hundreds of micrometers, and it could be developed in basic chemical labs without complex equipment.

Author Contributions: O.P.-D. made the experiments, performed SEM microscopy, and wrote the first manuscript draft. E.Q.-G. proposed and coordinated the project, performed the analysis of the results, and corrected the manuscript. All authors have read and agreed to the published version of the manuscript.

Funding: This work was funded by Projects CONACyT CB-2014-01-243407 and PROMEP BUAP-NPTC-377.

Conflicts of Interest: The authors declare no conflict of interest.

References

1. González-Fernández, A.; Juvert, J.; Aceves-Mijares, M.; Domínguez, C. Monolithic integration of a silicon-based photonic transceiver in a CMOS process. *IEEE Photonics J.* **2016**, *8*. [CrossRef]
2. Quiroga-González, E.; Carstensen, J.; Föll, H. Scalable processing and capacity of Si microwire array anodes for Li ion batteries. *Nanoscale Res. Lett.* **2014**, *9*, 417–421. [CrossRef] [PubMed]
3. Schöning, M.J.; Ronkel, F.; Crott, M.; Thust, M.; Schultze, J.W.; Kordos, P.; Lüth, H. Miniaturization of potentiometric sensors using porous silicon microtechnology. *Electrochim. Acta* **1997**, *42*, 3185–3193. [CrossRef]
4. Hildreth, O.; Lin, W.; Wong, C. Effect of catalyst shape and etchant composition on etching direction in metal-assisted chemical etching of silicon to fabricate 3D nanostructures. *ACS Nano* **2009**, *3*, 4033–4042. [CrossRef] [PubMed]
5. Kim, S.; Khang, D. Bulk micromachining of Si by metal-assisted chemical etching. *Small* **2014**, *10*, 3761–3766. [CrossRef] [PubMed]
6. Hildreth, O.; Brown, D.; Wong, C. 3D Out-of-plane rotational etching with pinned catalysts in metal-assisted chemical etching of silicon. *Adv. Funct. Mater.* **2011**, *21*, 3119–3128. [CrossRef]

7. Rezvani, S.J.; Pinto, N.; Boarino, L. Rapid formation of single crystalline Ge nanowires by anodic metal assisted etching. *CrystEngComm* **2016**, *18*, 7843–7848. [CrossRef]

8. Wilhelm, T.S.; Soule, C.W.; Baboli, M.A.; O'Connell, C.J.; Mohseni, P.K. Fabrication of Suspended III–V Nanofoils by Inverse Metal-Assisted Chemical Etching of In0.49Ga0.51P/GaAs Heteroepitaxial Films. *ACS Appl. Mater. Interfaces* **2018**, *10*, 2058–2066. [CrossRef]

9. DeJarld, M.; Shin, J.C.; Chern, W.; Chanda, D.; Balasundaram, K.; Rogers, J.A.; Li, X. Formation of High Aspect Ratio GaAs Nanostructures with Metal-Assisted Chemical Etching. *Nano Lett.* **2011**, *11*, 5259–5263. [CrossRef]

10. Yae, S.; Morii, Y.; Fukumuro, N.; Matsuda, H. Catalytic activity of noble metals for metal-assisted chemical etching of silicon. *Nanoscale Res. Lett.* **2012**, *7*, 352. [CrossRef]

11. Tsujino, K.; Matsumura, M. Morphology of nanoholes formed in silicon by wet etching in solutions containing HF and H_2O_2 at different concentrations using silver nanoparticles as catalysts. *Electrochim. Acta* **2007**, *53*, 28–34. [CrossRef]

12. Hadjersi, T.; Gabouze, N.; Kooij, E.S.; Zinine, A.; Ababou, A.; Chergui, W.; Cheraga, H.; Belhousse, S.; Djeghri, A. Metal-assisted chemical etching in $HF/Na_2S_2O_8$ or $HF/KMnO_4$ produces porous silicon. *Thin Solid Films* **2008**, *459*, 271–275. [CrossRef]

13. Han, H.; Huang, Z.; Lee, W. Metal-assisted chemical etching of silicon and nanotechnology applications. *Nano Today* **2014**, *9*, 271–304. [CrossRef]

14. Li, S.; Ma, W.; Zhou, Y.; Chen, X.; Xiao, Y.; Ma, M.; Zhu, W.; Wie, F. Fabrication of porous silicon nanowires by MACE method in $HF/H_2O_2/AgNO_3$ system at room temperature. *Nanoscale Res. Lett.* **2014**, *9*, 196–203. [CrossRef]

15. Peng, K.; Lu, A.; Zhang, R.; Lee, S.T. Motility of metal nanoparticles in silicon and induced anisotropic silicon etching. *Adv. Funct. Mater.* **2008**, *18*, 3026–3035. [CrossRef]

16. Pérez-Díaz, O.; Quiroga-González, E.; Silva-González, N.R. Silicon microstructures through the production of silicon nanowires by metal-assisted chemical etching, used as sacrificial material. *J. Mater. Sci.* **2019**, *54*, 2351–2357. [CrossRef]

17. Kolasinski, K.W.; Unger, B.A.; Ernst, A.T.; Aindow, M. Crystallographically determined etching and its relevance to the Metal-Assisted Catalytic Etching (MACE) of silicon powders. *Front. Chem.* **2019**, *6*, 651. [CrossRef]

18. Spira, M.E.; Hai, A. Multi-electrode array technologies for neuroscience and cardiology. *Nat. Nanotechnol.* **2013**, *8*, 83–94. [CrossRef]

19. Polyakov, D.S.; Veiko, V.P.; Salnikov, N.M.; Shimko, A.A.; Mikhaylova, A.A. Formation of microconical structures on Si surface under the action of pulsed periodic Yb-fiber laser radiation. *J. Phys. Conf. Ser.* **2018**, *951*, 012012. [CrossRef]

20. Korany, F.M.H.; Hameed, M.F.O.; Hussein, M.; Mubarak, R.; Eladawy, M.I.; Obayya, S.S.A. Conical structures for highly efficient solar cell applications. *J. Nanophotonics* **2018**, *12*, 016019. [CrossRef]

21. Schneider, L.; Laustsen, M.; Mandsberg, N.; Taboryski, R. The influence of structure heights and opening angles of micro- and nanocones on the macroscopic surface wetting properties. *Sci. Rep.* **2016**, *6*, 21400. [CrossRef]

22. Ding, W.; Fernandino, M.; Dorao, C.A. Conical micro-structures as a route for achieving super-repellency in surfaces with intrinsic hydrophobic properties. *Appl. Phys. Lett.* **2019**, *115*, 053703. [CrossRef]

23. Huang, Z.; Fang, H.; Zhu, J. Fabrication of silicon nanowire arrays with controlled diameter, length, and density. *Adv. Mater.* **2007**, *19*, 744–748. [CrossRef]

24. Huang, Z.; Shimizu, T.; Senz, S.; Zhang, Z.; Zhang, X.; Lee, W.; Geyer, N.; Gösele, U. Ordered arrays of vertically aligned [110] silicon nanowires by suppressing the crystallographically preferred <100> etching directions. *Nano Lett.* **2009**, *9*, 2519–2525. [CrossRef] [PubMed]

25. Chang, S.W.; Chuang, V.; Boles, S.; Ross, C.; Thompson, C. Densely packed arrays of ultra-high-aspect-ratio silicon nanowires fabricated using block-copolymer lithography and metal-assisted etching. *Adv. Funct. Mater.* **2009**, *19*, 2495–2500. [CrossRef]

26. Peng, K.; Hu, J.; Yan, Y.; Wu, Y.; Fang, H.; Xu, Y.; Lee, S.; Zhu, J. Fabrication of single-crystalline silicon nanowires by scratching a silicon surface with catalytic metal particles. *Adv. Funct. Mater.* **2006**, *16*, 387–394. [CrossRef]

27. Chartier, C.; Bastide, S.; Lévy-Clément, C. Metal-assisted chemical etching of silicon in HF–H_2O_2. *Electrochim. Acta* **2008**, *53*, 5509–5516. [CrossRef]

28. Yun, M. Investigation of KOH anisotropic etching for the fabrication of sharp tips in silicon-on-insulator (soi) material. *J. Korean Phys. Soc.* **2000**, *37*, 605–610. [CrossRef]

29. Seidel, H.; Csepregi, L.; Heuberger, A.; Baumgärtel, H. Anisotropic etching of crystalline silicon in alkaline Solutions. *J. Electrochem. Soc.* **1990**, *137*, 3612–3626. [CrossRef]

30. Nguyen, N.-T. Chapter 4. Fabrication technologies. *Micromixers* **2012**, 113–161. [CrossRef]

31. Zhang, X.G. *Electrochemistry of Silicon and Its Oxide*; Kluwer Academic Publishers: New York, NY, USA, 2001; p. 134, ISBN 0306479214.

32. Quiroga-González, E.; Juárez-Estrada, M.A.; Gómez-Barojas, E. Light Enhanced Metal Assisted Chemical Etching of Silicon. *ECS Trans.* **2018**, *86*, 55–63. [CrossRef]

33. Um, H.-D.; Kim, N.; Lee, K.; Hwang, I.; Seo, J.H.; Yu, Y.J.; Duane, P.; Wober, M.; Seo, K. Versatile control of metal-assisted chemical etching for vertical silicon microwire arrays and their photovoltaic applications. *Sci. Rep.* **2015**, *5*, 11277. [CrossRef] [PubMed]

© 2020 by the authors. Licensee MDPI, Basel, Switzerland. This article is an open access article distributed under the terms and conditions of the Creative Commons Attribution (CC BY) license (http://creativecommons.org/licenses/by/4.0/).

 *micromachines* 

Article

Fabrication of Ultra-High Aspect Ratio (>420:1) Al$_2$O$_3$ Nanotube Arraysby Sidewall TransferMetal Assistant Chemical Etching

Hailiang Li and Changqing Xie *

Key Laboratory of Microelectronic Devices & Integrated Technology, Institute of Microelectronics of Chinese Academy of Sciences, Beijing 100029, China; lihailiang@ime.ac.cn
* Correspondence: xiechangqing@ime.ac.cn; Tel.: +86-01-82995581

Received: 28 February 2020; Accepted: 30 March 2020; Published: 3 April 2020

Abstract: We report a robust, sidewall transfer metal assistant chemical etching scheme for fabricating Al$_2$O$_3$ nanotube arrays with an ultra-high aspect ratio. Electron beam lithography followed by low-temperature Au metal assisted chemical etching (MacEtch) is used to pattern high resolution, high aspect ratio, and vertical silicon nanostructures, used as a template. This template is subsequently transferred by an atomic layer deposition of the Al$_2$O$_3$ layer, followed by an annealing process, anisotropic dry etching of the Al$_2$O$_3$ layer, and a sacrificial silicon template. The process and characterization of the Al$_2$O$_3$ nanotube arrays are discussed in detail. Vertical Al$_2$O$_3$ nanotube arrays with line widths as small as 50 nm, heights of up to 21 µm, and aspect ratios up to 420:1 are fabricated on top of a silicon substrate. More importantly, such a sidewall transfer MacEtch approach is compatible with well-established silicon planar processes, and has the benefits of having a fully controllable linewidth and height, high reproducibility, and flexible design, making it attractive for a broad range of practical applications.

Keywords: Al$_2$O$_3$ nanotube; ultra-high aspect ratio; gold (Au) metal assisted chemical etching; atomic layer deposition; anisotropic dry etching

---

1. Introduction

In recent years, Al$_2$O$_3$ nanotube arrays possessing high surface-to-volume ratios have attracted much attention, owing to their potential applications in optoelectronics [1,2], biotechnology, and photocatalysis [3,4]. For example, several recent studies have indicated that Al$_2$O$_3$ nanotube arrays exhibit excellent dielectric properties [5] and good flexibility [6], as compared with other oxide nanotubes. Thus, Al$_2$O$_3$ nanotube arrays would be more advantageous for use as optical transportation media in optoelectronics. They can also be utilized as a new biomineralization nanoreactor in biotechnology, and so on. Various template-based strategies for the fabrication of Al$_2$O$_3$ nanotubes with a high aspect ratio have been proposed. These include hydrothermal reaction methods [7], coating of carbon nanotubes with aluminum isopropoxide [8], anodization of Si-based Al film [9,10], and etching of a porous anodic alumina (PAA) template in a NaOH solution and ultrasonic treatment of PAA membrane [11,12]. In addition to the Al$_2$O$_3$ nanotubes, atomic layer deposition (ALD) coatings of silicon oxide (SiO$_2$) [13], zinc oxide (ZnO) [14,15], zirconium dioxide(ZrO$_2$) [16], and titanium oxide(TiO$_2$) [16,17] have also been successfully applied to the preparation of nanostructures.Using these methods, much progress has been made on achieving highly ordered nanotube arrays with a controllable size and short length. However, due to the poor consistency of the pillar size in the templates, it is not easy to confine and control the profile of the fabricated nanostructures. Moreover, separating hollow nanostructures from a template to obtain individual nanostructures remains a challenge. Many kinds of nanotubes with high aspect ratios will lose their tubular structure after extraction from the template. With the continuous request for high performance nanotube-based devices, the aspect ratio of the fabricated nanotube arrays as obtained by tailoring electrochemical conditions is far from satisfactory.

Metal assisted chemical etching (MacEtch), first reported by Li and Bohn in 2001 [18], offers a promising wet etching solution for generating silicon nanostructures [19–22]. In addition, the MacEtch of germanium (Ge) [23] and III–V compound semiconductors, such as GaN [24,25], GaAs [26–28], GaP [29], InP [30], AlGaAs [31], InGaAs [32], and InGaP [27], have also been demonstrated. This method has the benefits of inherent simplicity, a low cost, high versatility, and high reproducibility, making it attractive for preparing silicon nanowire arrays. The conventional method for generating micrometer, submicrometer, and nanosized silicon structures with high aspect ratios [19–22] is by submerging a metal-coated Si sample into an etchant solution composed of hydrofluoric acid, hydrogen peroxide solution, and a diluting agent. Furthermore, large-area uniform micro-gratings with well controlled morphological features and depths as large as 80 μm have also been successfully produced by optimizing the MacEtch method [33]. Highly dense Si/TiO_2 core/shell nanowire arrays have also been synthesized using a nanostructured Si template obtained by the MacEtch of the Si substrates and a layer of TiO_2 deposited by ALD [34,35]. In the previous works, using Ti/Au nanostructures patterned with electron-beam lithography followed by ion beam etching, we fabricated vertical silicon nanopillar arrays with a period of 250 nm and an aspect ratio of 160:1 using MacEtch [36]. A 20 nm minimum feature size was also realized by MacEtch-based nanoimprinting [37]. Unfortunately, the MacEtch method cannot be directly adopted to generate oxide nanotubes.

Sidewall transfer lithography has been widely recognized as a promising technology that can fabricate deca-nanometer metal-oxide-semiconductor field-effect transistors (MOSFETs) and siliconnanostructures [38–40]. The key idea is to combine three well-established techniques (lithography, sidewall process, and dry etching) to create silicon nanostructures. In this work, we present a reliable means, called sidewall transfer metal assistant chemical etching, to fabricate Al_2O_3 nanotube arrays with an ultra-high aspect ratio. The key idea is to combine a low-temperature MacEtch and sidewall transfer process based on ALD and dry etching to simultaneously achieve their respective advantages. The former is used to generate vertically oriented silicon nanostructures that serve as a sacrificial layer, while the latteris used to transfer the silicon nanostructures to Al_2O_3 nanotube arrays with a higher aspect ratio and smaller feature size. The ultimate aspect ratio and feature size of the fabricated Al_2O_3 nanotube arrays can be controlled by modifying the thickness of the Al_2O_3 film deposited by ALD. The influence of the deposition temperature and annealing temperature on the structure and optical properties of Al_2O_3 thin films deposited by ALD are also examined.

2. Materials and Methods

The process flow of the proposed sidewall transfer MacEtchis shown in Figure 1. In our experiments, 4in.p-Type<100>CZsiliconwafers (Silicon Quest International, San Jose, CA, USA) with a resistivity of 2–10 Ω·cm and thickness of 500 μm were used. After the samples (Figure 1a) were cleaned in sulfuric acid and hydrogen peroxide to remove the surface native oxide layers, a 3-nmTi/20 nm Au thin film was deposited on the silicon substrates by a magnetron sputtering system (ACS-4000, ULVAC Company, Chigasaki, Japan). The working pressure was maintained at 4.5×10^{-6} Torr, and the temperature of the chamber was kept at 25 °C over the entire deposition process. Then, highly sensitive chlorinated electron beam resist ZEP520A was spin-coated on the Ti/Au layer to a thickness of about 400 nm, and was bakedon a hotplate at 180 °C for 2 min (Figure 1b). The resist was exposed with an electron beam lithography system (JBX-6300FS, JEOL Company, Tokyo, Japan) for patterning definition (Figure 1c), and ion beam etching (IBE) was performed to transfer ZEP520A resist pattern onto the Ti/Au layer, forming Ti/Au nanostructures that serve as a local cathode (Figure 1d). Then, the MacEtch processwas carried out to generate silicon nanostructures with a high aspect ratio (Figure 1e). The ALD process was used to deposit the Al_2O_3 film on the sidewalls and on top of the generated silicon nanostructures (Figure 1g). Finally, the exposed tops of the Al_2O_3 film and its silicon nanostructures underneath were removed by dry etching, forming Al_2O_3 nanotube arrays with a higher aspect ratio and smaller feature size.

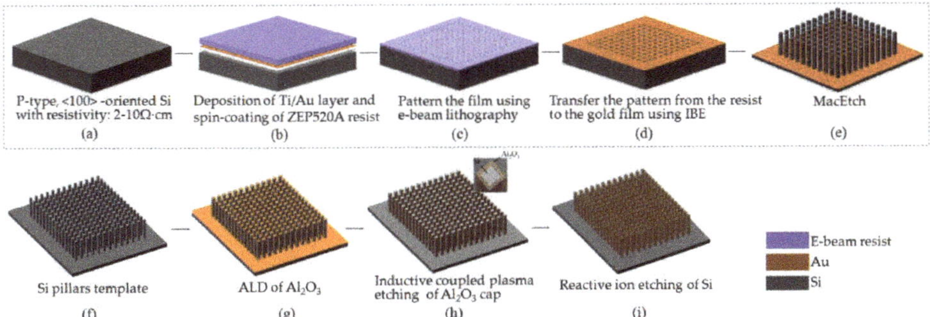

Figure 1. Schematic diagram of the sidewall transfer MacEtch process for the realization of the Al_2O_3 nanotube arrays with ultra-high aspect ratios: (**a**) silicon substrate; (**b**) deposition of the Ti/Au layer and ZEP520A resist; (**c**) electron beam lithography; (**d**) ion beam etching; (**e**) metal assisted chemical etching (MacEtch); (**f**) Si pillars template; (**g**) atomic layer deposition (ALD) Al_2O_3 film; (**h**) dry etching; (**i**) reactive ion etching of Si.

The morphology and structure of the fabricated Al_2O_3 nanotube arrays were characterized by a scanning electron microscope (SEM; SUPRA-55, Zeiss, Oberkochen, Germany). The information on the surface chemistry of the Al_2O_3 layer deposited by the atomic layer deposition (ALD) technique was investigated using X-ray photo emission spectroscopy (XPS) with a monochromatic Al Kα X-ray source. Both the Al_2O_3 film thickness and its refractive index profile were determined by a spectroscopic ellipsometry (Horiba Uvisel FUV, Kyoto, Japan) over the spectral range of 150 to 900 nm using the Cauchy model.

2.1. Fabrication of Ti/Au Nanostructures with Low Aspect Ratio

To obtain high-resolution resist patterns, electron beam lithography was performed at an accelerating voltage of 100 kV, with a beam current of 200 pA and an exposure dose of 400 $\mu C/cm^2$. After electron beam exposure, the ZEP520A resist was developed using a standard developer N50D (ZEON) for 1 min at 18 °C, and rinsed with isopropanol (IPA) for 40 s to stop development. Then, the sample was dried with a steady stream of N_2.

Next, a home-made argon (Ar) IBE system was used to transfer the resist patterns into a Ti/Au thin film deposited onto the silicon wafer. The working pressure was maintained at 1.0×10^{-4} Torr, and the substrate temperature was lower than 100 °C over the entire IBE process. The Ar^+ ion energy and the beam current density were 500 eV and 1 mA/cm^2, respectively. The corresponding etching rates of Au and the resist were 120 nm/min and 20 nm/min, respectively, resulting in a selectivity ratio of 6:1. The etching time in our experiments was 12 s. After completion of the IBE process, the residual ZEP520A resist was removed by washing with a resist removal solution (ZDMAC, ZEON Company, Tokyo, Japan), followed by oxygen plasma ashing. The patterned Ti/Au nanostructures served as a catalyst in the subsequent low-temperature MacEtch process.

2.2. Fabrication of Silicon Nanostructures with High Aspect Ratio

A low-temperature Au MacEtch process was performed to generate silicon nanostructures with a high aspect ratio. It should be noted that samples must be kept clean and tidy before the Au MacEtch process. The Au MacEtch process was carried out in an etchant solution composed of hydrofluoric acid, hydrogen peroxide solution, and deionized water (4.1 M HF/0.15 M H_2O_2/45 M H_2O) at 2 °C. The etching was conducted for 15 min. It is well known that the collapse of large aspect ratio nanostructures often occurs during the drying step. In our drying scheme, we used isopropanol with low surface tension, instead of deionized water (DI) water as a rinse solution. This reduced the capillary force acting on the silicon nanostructures. After the sample was rinsed with isopropanol, it was evaporated naturally and dried at room temperature.

2.3. Al_2O_3 Film Deposition by ALD

The ALD process allows one to precisely deposit highly uniform and conformal thin films onto complex three-dimensional topographies. Here, an Al_2O_3 film was deposited using $Al(CH_3)_3$ (TMA) and deionized water with a hot-wall atomic layer deposition system (Picosun R200, Espoo, Finland). The deposition temperature was 300 °C. An Al_2O_3 film was grown on the sidewalls and on top of the fabricated silicon nanostructures with an aspect ratio of 160:1, using TMA and deionized water. TMA (Sigma Aldrich, St. Louis, MO, USA) was used as the precursor and deionized water was used as an oxidant source during the ALD process. The TMA reactant exposure time, N_2 purge time following TMA reactant exposure, water exposure time, and N_2 purge time following the H_2O reactant exposure were 0.5 s, 2 s, 0.5 s, and 2 s, respectively. The growth rate of the Al_2O_3 film was 0.089 nm/cycle, as inferred by spectroscopic ellipsometry. The sample was processed with 400 ALD cycles, and the corresponding Al_2O_3 film thickness was 50 nm.

2.4. Annealing Processand Characterization of the Al_2O_3 Film

The thermal expansion coefficient of Al_2O_3 film is 8.2×10^{-6} °C^{-1}, which is higher than that of silicon (2.6×10^{-6} °C^{-1}). To compensate the lattice mismatch between the Al_2O_3 ALD filmand silicon nanostructures, four samples were annealed at 700 °C, 800 °C, 900 °C, and 1000 °C under a vacuum for 90 min, seperately. The specifications of the Al_2O_3 film was investigated using X-ray photoelectron spectra (XPS). Characterization of the optical properties of our Al_2O_3 film from 150 to 900 nm wavelengths was also performed using spectroscopic ellipsometry.

2.5. Formation of Al_2O_3 Nanotube Arrays with an Ultra-High Aspect Ratio

First, the exposed tops of the Al_2O_3 film on the silicon nanostructures were removed by an inductive coupled plasma (ICP) etching system (ULVAC, Japan), with a mixture of BCl_3 and Cl_2 reactive gas. Secondly, the silicon nanostructures underneath the exposed tops of the Al_2O_3 film were removed by the same ICP etching system with SF_6 reactive gas and an isotropic reactive ion etching mode. The etching parameters of the Al_2O_3 film and silicon nanostructuresa re summarized in Table 1.

Table 1. Recipes for Al_2O_3 and Si in etch system.

Process Parameters	Al_2O_3 Etch	Si Etch
Cl_2 (sccm)	1.2	–
BCl_3 (sccm)	6.8	–
SF_6 (sccm)	–	90
Pressure (mtorr)	3	4
Coil power (W)	900	400
Platen power (W)	200	3

Process temperature is 25 °C for all of the processes.

The etch rates of the Al_2O_3 film and silicon nanostructures were 0.89 nm/s and 9 nm/s, respectively. The etching selectivity between the silicon nanostructures and Al_2O_3 film was as high as 66,000:1 for the Si etching recipe [41].

3. Results and Discussion

Figure 2 shows the dependence of the Au MacEtch process parameters (etchant solutions and etching temperature) on the quality of the fabricated silicon nanostructures. When the hydrogen peroxide concentration in the etchant solution was relatively high, the lateral etching rate increased and defects began to occur on the sidewalls, as shown in Figure 2a. This is because as the concentration of H_2O_2 increased, the number of the generated holes also increased, resulting in an increase in the silicon etching rate. In other words, when the generated holes between the Si and Ti/Au interfaces could not be completely consumed, the excess holes spread laterally, leading to lateral corrosion and the formation of defects in the sidewalls. Using an optimized H_2O_2 concentration, the transverse etching rate could be limited by the availability of the generated hole. Thus, vertical silicon nanostructures could be obtained, as shown in Figure 2b.

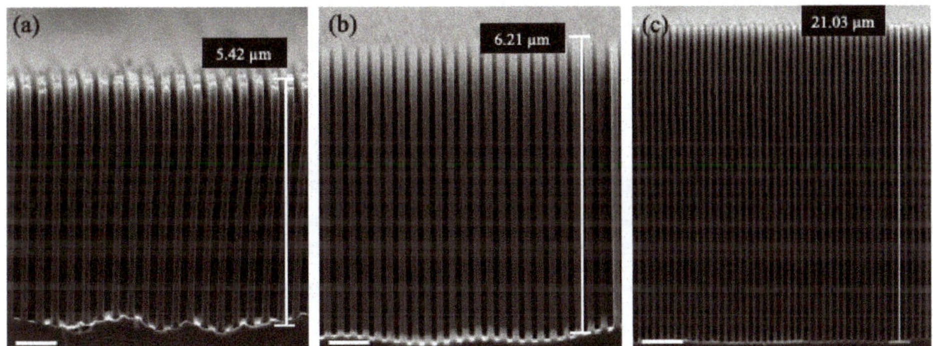

Figure 2. Cross-sectional scanning electron microscope (SEM) micrographs of silicon nanostructures using various etchant solutions composed of hydrofluoric acid, hydrogen peroxide solution, and deionised (DI) water with molar ratios of (**a**) 4.8:0.35:50, (**b**) 4.8:0.15:50, and (**c**) 4.8:0.15:50. The scale bars in (**a**–**c**) are 850 nm, 900 nm, and 1.9 µm, respectively. The etching temperatures in (**a**,**b**) are 22 °C, and the etching temperature in (**c**) is 2 °C.

Etching temperature also plays an essential role in the Au MacEtch process. When the Au MacEtch process was performed at 2 °C, a better morphology could be obtained than that at room temperature, at the expense of a much lower etching rate. In this work, a low-temperature Au MacEtch process could result in silicon nanostructures with an aspect ratio of up to 160:1, as shown in Figure 2c. Silicon nanostructure bottoms that are especially clean and flat can be obtained.

Figure 3 shows the binding energies of the Al_2O_3 film deposited at 300 °C. Two signatures of orbital, 74.4 eV for Al (2p) and 531.5 eV for O (1s), can be observed from an XPS wide scan. The O/Al ratio of the Al_2O_3 film is close to the expected value of 1.5, corresponding to the lattice oxygen of Al_2O_3 [42]. The difference between these two elemental peaksis close to the standard values in the literature for different forms of aluminum oxide [43,44]. A peak of binding energy of 1s carbon in the Al_2O_3 film could also be observed, indicating either an incomplete reaction or an insufficient N_2 purge time.

Figure 4 shows the Al_2O_3 ALD film thickness and refractive index as a function of the annealing temperature. The refractive index of the Al_2O_3 ALD filminitially increased with an increase of annealing temperature from 700 °C to 800 °C, and later saturated with the increasing annealing temperature. When the annealing temperature reached 1150 °C, the refractive index approached a maximized value of 1.724, which is slightly smaller than that of crystalline sapphire (1.76). The measured data indicate that the amorphous Al_2O_3 ALD film in this experiment would densify further upon crystallization. This is consistent with the existing result [45]. Meanwhile, there was a 10% decrease in the Al_2O_3 ALD film thickness with increasing the annealing temperature from 700 °C to 800 °C, which was mainly due to an increase inthe density and purity levels of the Al_2O_3 films deposited by ALD. The measured data also indicate the occurrence of Al_2O_3 crystallization after high-temperature annealing, which appears to be particularly advantageous for preventing the collapse of ultra-high aspect ratio Al_2O_3 nanotube arrays.

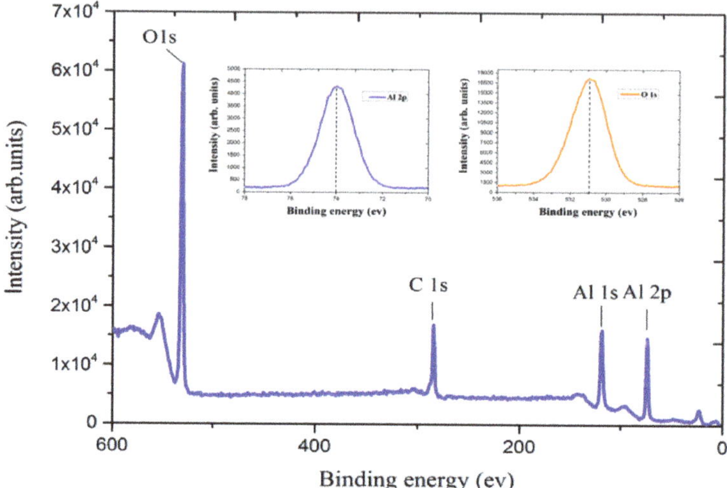

Figure 3. X-ray photoelectron spectra of the Al_2O_3 layer deposited at 300 °C. The two insets show the Al (2p) and O (1s) peaks, respectively.

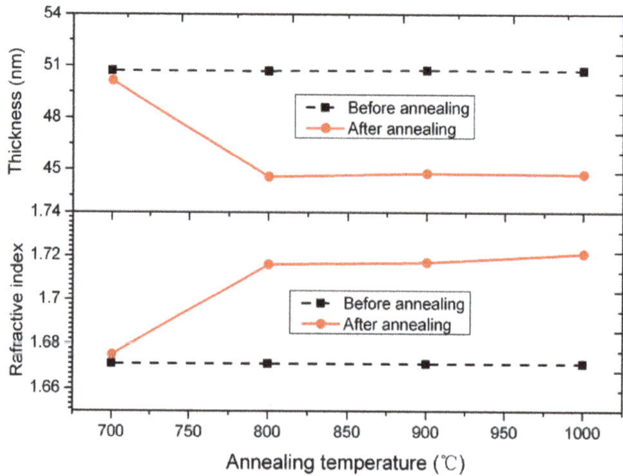

Figure 4. Variation of film thickness and refractive index versus annealing temperature.

Figure 5a shows the SEM image of the top view of the generated Si nanostructures, followed by the Al_2O_3 film deposition by ALD and annealedat 1150 °C for 90 min. The results of the generated Al_2O_3 nanotube arrays without the annealing process are also given in Figure 5b. One can clearly see that the Al_2O_3 nanotube arrays bend and lean against each other.The bending is probably caused by a mismatch in internal stress between the Al_2O_3 layer and Si pillar. The thermal expansion coefficient of the Al_2O_3 film is 8.2×10^{-6} °C^{-1}, while the thermal expansion coefficient of silicon is 2.6×10^{-6} °C^{-1}. It should be noted that the Al_2O_3 ALD film was deposited by the temperature gradient method over a temperature range from room temperature to 300 °C, and the Si cores were removed afterwards using a plasma reactive ion etching at 25 °C. By comparison, Figure 5c,d shows the SEM images of the top andcross section views, respectively, of the generated Al_2O_3 nanotube arrays after the whole process was finished. These uniform structures had a line width of 50 nm and height of 21 µm, corresponding to an aspect ratio as high as 420:1. The Al_2O_3 nanotube arrays were perfectly preserved from collapsing after the annealing process, confirming the successful pattern transfer of the sidewall transfer MacEtch process. The shape of the Al_2O_3 nanotube arrays exhibited very little deformation relative to that of the original Si nanostructures. Figure 5e shows a tilted SEM view of the Al_2O_3 nanotube arrays after the whole process. One can see that the silicon template was almost completely removed by plasma reactive ion etching at 25 °C.

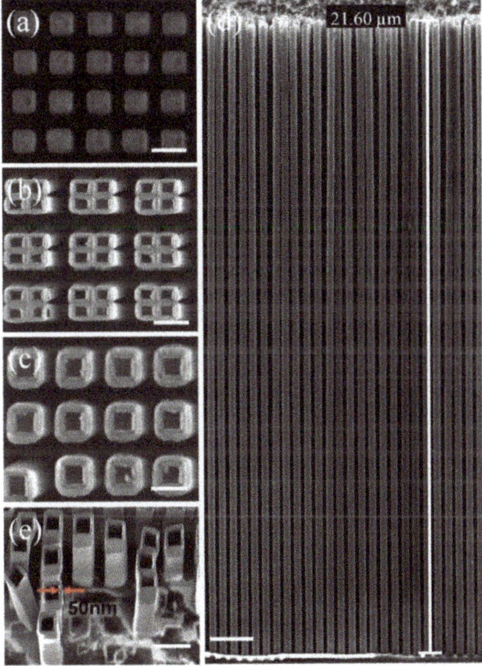

Figure 5. A series of images demonstrating sidewall transfer MacEtch for the preparation of ultra-high aspect ratio Al_2O_3 nanotube arrays. (**a**) Top view SEM of the Si nanostructures after Al_2O_3 film deposition by ALD. The scale bar is 300 nm. (**b**) Top view SEM of the generated Al_2O_3 nanotube arrays without annealing process. The scale bar is 400 nm. (**c,d**) SEM images of the top and cross section views of the generated Al_2O_3 nanotube arrays resulting from the sidewall transfer MacEtch process. The scale barsin (**c,d**) are 250 nm and 1.6 µm, respectively. (**e**) Tilted SEM view of Al_2O_3 nano-tube structures after the whole process. The scale bar is 400 nm. The tilting angle is 35°.

4. Conclusions

In summary, we have demonstrated a reliable sidewall transfer MacEtch process to fabricate ultra-high aspect ratio Al_2O_3 nanotube arrays with linewidthsas small as 50 nm, heights of up to 21 µm, and an aspect ratio of up to 420:1. This technique combines the advantages of the high aspect ratio nanostructure capabilities of the low-temperature Au MacEtch with thesidewall transfer process. The use of the sidewall transfer process has two advantages. First, it leads to higher-resolution and higher aspect ratio Al_2O_3 nanotube patterns than with MacEtch, and second, the line-width of the Al_2O_3 nanotubes can be precisely controlled by the cycle number of ALD for Al_2O_3. The sidewall transfer MacEtch provides a promising route to the scalable manufacturing of ultra-high aspect ratio

Al$_2$O$_3$ nanotube arrays, and is applicable to other kinds of oxide nanotubes, as long as the oxide can be deposited by ALD.

Author Contributions: H.L. and C.X. conceived the idea for the experiment; H.L. performed the experiments; H.L. and C.X. wrote the manuscript. All authors have read and agreed to the published version of the manuscript.

Funding: This work was supported by the National Key Research and DevelopmentProgram of China (grant no. 2017YFA0206002) and the National Natural Science Foundation of China (grant no. U1832217, no. 61804169, and no. 61821091).

Conflicts of Interest: The authors declare no conflict of interest.

References

1. Shanmugham, S.; Hendrick, M.; Richards, N.; Oljaca, M. Candidate oxidation resistant coatings via combustion chemical vapor deposition. *J. Mater. Sci.* **2004**, *39*, 377–381. [CrossRef]

2. Gutiérrez, M.; Lloret, F.; Pham, T.T.; Cañas, J.; Reyes, D.F.D.L.; Eon, D.; Pernot, J.; Araujo, D. Control of the Alumina Microstructure to Reduce Gate Leaks in Diamond MOSFETs. *Nanomater* **2018**, *8*, 584. [CrossRef] [PubMed]

3. Thamaraiselvi, T.V.; Rajewari, S. Biological evaluation of bioceramic materials—Areview. *Trends Biomater. Artif. Organs* **2004**, *81*, 9–17.

4. Xifré-Pérez, E.; Ferre-Borull, J.; Pallarès, J.; Marsal, L.F. Mesoporous alumina as a biomaterial for biomedical applications. *Open Mater. Sci.* **2015**, *2*, 13–32. [CrossRef]

5. De Wit, H.; Crevecoeur, C. The dielectric breakdown of anodic aluminum oxide. *Phys. Lett. A* **1974**, *50*, 365–366. [CrossRef]

6. Mei, Y.; Wu, X.; Shao, X.F.; Siu, G.G.; Bao, X.M. Formation of an array of isolated alumina nanotubes. *EPL Europhys. Lett.* **2003**, *62*, 595–599. [CrossRef]

7. Lee, H.C.; Kim, H.J.; Chung, S.H.; Lee, K.H.; Lee, H.C.; Lee, J.S. Synthesis of unidirectional alumina nanostructures without added organics olvents. *J. Am. Chem. Soc.* **2003**, *125*, 2882–2883. [CrossRef] [PubMed]

8. Zhang, Y.; Liu, J.; He, R.; Zhang, Q.; Zhang, X.; Zhu, J. Synthesis of alumina nanotubes using carbon nanotubes as templates. *Chem. Phys. Lett.* **2002**, *360*, 579–584. [CrossRef]

9. Mei, Y.; Wu, X.; Shao, X.; Huang, G.; Siu, G. Formation mechanism of alumina nanotube array. *Phys. Lett. A* **2003**, *309*, 109–113. [CrossRef]

10. Pu, L.; Bao, X.M.; Zou, J.P.; Feng, D. Individual alumina nanotubes. *Angew. Chem. Int. Ed.* **2001**, *113*, 1538. [CrossRef]

11. Xiao, Z.-L.; Han, C.; Welp, U.; Wang, H.H.; Vlasko-Vlasov, V.; Kwok, W.K.; Miller, D.J.; Hiller, J.; Cook, R.E.; Willing, G.A.; et al. Nickel antidot arrays on anodic alumina substrates. *Appl. Phys. Lett.* **2002**, *81*, 2869–2871. [CrossRef]

12. Huang, G.S.; Wu, X.L.; Xie, Y.; Shao, X.F.; Wang, S.H. Light emission from silicon-based porous anodic alumina form edin 0.5 M oxalicacid. *J. Appl. Phys.* **2003**, *94*, 2407–2410. [CrossRef]

13. Jensen, D.S.; Kanyal, S.S.; Madaan, N.; Miles, A.J.; Davis, R.C.; Vanfleet, R.; Vail, M.A.; Dadson, A.E.; Linford, M.R. Ozone priming of patterned carbon nanotube forests for subsequent atomic layer deposition-like deposition of SiO$_2$ for the preparation of microfabricated thin layer chromatography plates. *J. Vac. Sci. Technol. B* **2013**, *31*, 31803. [CrossRef]

14. Zhang, Y.; Liu, M.; Ren, W.; Ye, Z.-G. Well-ordered ZnO nanotube arrays and networks grown by atomic layer deposition. *Appl. Surf. Sci.* **2015**, *340*, 120–125. [CrossRef]

15. Elam, J.W.; Routkevitch, D.; Mardilovich, P.P.; George, S.M. Conformal coating on ultra high-aspect-ratio nanopores of anodic alumina by atomic layer deposition. *Chem. Mater.* **2003**, *15*, 3507–3517. [CrossRef]

16. Zazpe, R.; Knaut, M.; Sopha, H.; Hromadko, L.; Albert, M.; Prikryl, J.; Gärtnerová, V.; Bartha, J.W.; Macak, J.M. Atomic Layer Deposition for Coating of High Aspect Ratio TiO$_2$ Nanotube Layers. *Langmuir* **2016**, *32*, 10551–10558. [CrossRef]

17. Shin, H.; Jeong, D.-K.; Lee, J.; Sung, M.M.; Kim, J. Formation of TiO$_2$ and ZrO$_2$ Nanotubes Using Atomic Layer Deposition with Ultraprecise Control of the Wall Thickness. *Adv. Mater.* **2004**, *16*, 1197–1200. [CrossRef]

18. Li, X.; Bohn, P.W. Metal-assisted chemical etching in HF/H$_2$O$_2$ produces porous silicon. *Appl. Phys. Lett.* **2000**, *77*, 2572–2574. [CrossRef]
19. Bai, S.; Du, Y.; Wang, C.; Wu, J.; Sugioka, K. Reusable Surface-Enhanced Raman Spectroscopy Substrates Made of Silicon Nanowire Array Coated with Silver Nanoparticles Fabricated by Metal-Assisted Chemical Etching and Photonic Reduction. *Nanomater* **2019**, *9*, 1531. [CrossRef]
20. Chang, C.; Sakdinawat, A. Ultra-high aspect ratio high-resolution nanofabrication for hard X-ray diffractive optics. *Nat. Commun.* **2014**, *5*, 4243. [CrossRef] [PubMed]
21. Chen, X.; Bi, Q.; Sajjad, M.; Wang, X.; Ren, Y.; Zhou, X.; Xu, W.; Liu, Z. One-Dimensional Porous Silicon Nanowires with Large Surface Area for Fast Charge–Discharge Lithium-Ion Batteries. *Nanomaterials* **2018**, *8*, 285. [CrossRef] [PubMed]
22. Hoshian, S.; Gaspar, C.; Vasara, T.; Jahangiri, F.; Jokinen, V.; Franssila, S. Non-Lithographic Silicon Micromachining Using Inkjet and Chemical Etching. *Micromachines* **2016**, *7*, 222. [CrossRef]
23. Rezvani, S.J.; Pinto, N.; Boarino, L. Rapid formation of single crystalline Ge nanowires by anodic metal assisted etching. *CrystEngComm* **2016**, *18*, 7843–7848. [CrossRef]
24. Geng, X.; Duan, B.K.; Grismer, D.A.; Zhao, L.; Bohn, P.W. Monodisperse GaN nanowires prepared by metal-assisted chemical etching with in situ catalyst deposition. *Electrochem. Commun.* **2012**, *19*, 39–42. [CrossRef]
25. Duan, B.K.; Bohn, P.W. High sensitivity hydrogen sensing with Pt-decorated porous gallium nitride prepared by metal-assisted electroless etching. *Analyst* **2010**, *135*, 902. [CrossRef] [PubMed]
26. DeJarld, M.; Shin, J.C.; Chern, W.; Chanda, D.; Balasundaram, K.; Rogers, J.A.; Li, X. Formation of High Aspect Ratio GaAs Nanostructures with Metal-Assisted Chemical Etching. *Nano Lett.* **2011**, *11*, 5259–5263. [CrossRef]
27. Wilhelm, T.S.; Soule, C.W.; Baboli, M.A.; O'Connell, C.J.; Mohseni, P.K. Fabrication of Suspended III–V Nanofoils by Inverse Metal-Assisted Chemical Etching of In0.49Ga0.51P/GaAs Heteroepitaxial Films. *ACS Appl. Mater. Interfaces* **2018**, *10*, 2058–2066. [CrossRef]
28. Lova, P.; Robbiano, V.; Cacialli, F.; Comoretto, D.; Soci, C. Black GaAs by Metal-Assisted Chemical Etching. *ACS Appl. Mater. Interfaces* **2018**, *10*, 33434–33440. [CrossRef]
29. Kim, J.; Oh, J. Formation of GaP Nanocones and Micro-mesas by Metal-assisted Chemical Etching. *Phys. Chem. Chem. Phys.* **2016**, *18*, 3402–3408. [CrossRef]
30. Kim, S.H.; Mohseni, P.K.; Song, Y.; Ishihara, T.; Li, X. Inverse Metal-Assisted Chemical Etching Produces Smooth High Aspect Ratio InP Nanostructures. *Nano Lett.* **2014**, *15*, 641–648. [CrossRef]
31. Wilhelm, T.S.; Wang, Z.-H.; Baboli, M.A.; Yan, J.; Preble, S.F.; Mohseni, P.K. Ordered AlxGa1–xAs Nanopillar Arrays via Inverse Metal-Assisted Chemical Etching. *ACS Appl. Mater. Interfaces* **2018**, *10*, 27488–27497. [CrossRef] [PubMed]
32. Kong, L.; Song, Y.; Kim, J.D.; Yu, L.; Wasserman, D.; Chim, W.K.; Chiam, S.Y.; Li, X. Damage-Free Smooth-Sidewall InGaAs Nanopillar Array by Metal-Assisted Chemical Etching. *ACS Nano* **2017**, *11*, 10193–10205. [CrossRef]
33. Romano, L.; Vila-Comamala, J.; Jefimovs, K.; Stampanoni, M. Effect of isopropanol on gold assisted chemical etching of silicon microstructures. *Microelectron. Eng.* **2017**, *177*, 59–65. [CrossRef]
34. Scuderi, V.; Impellizzeri, G.; Romano, L.; Scuderi, M.; Nicotra, G.; Bergum, K.; Irrera, A.; Svensson, B.; Privitera, V. TiO$_2$-coated nanostructures for dye photo-degradation in water. *Nanoscale Res. Lett.* **2014**, *9*, 458. [CrossRef]
35. Hwang, Y.J.; Boukai, A.; Yang, P. High Density n-Si/n-TiO$_2$ Core/Shell Nanowire Arrays with Enhanced Photoactivity. *Nano Lett.* **2009**, *9*, 410–415. [CrossRef] [PubMed]
36. Li, H.; Ye, T.; Shi, L.; Xie, C. Fabrication of ultra-high aspect ratio (>160:1) silicon nanostructures by using Au metal assisted chemical etching. *J. Micromech. Microeng.* **2017**, *27*, 124002. [CrossRef]
37. Li, H.; Niu, J.; Wang, G.; Wang, E.; Xie, C. Direct Production of Silicon Nanostructures with Electrochemical Nanoimprinting. *ACS Appl. Electron. Mater.* **2019**, *1*, 1070–1075. [CrossRef]
38. Kim, D.H.; Sung, S.-K.; Sim, J.S.; Kim, K.R.; Lee, J.D.; Park, B.-G.; Choi, B.H.; Hwang, S.; Ahn, D. Single-electron transistor based on a silicon-on-insulator quantum wire fabricated by a side-wall patterning method. *Appl. Phys. Lett.* **2001**, *79*, 3812–3814. [CrossRef]
39. Liu, D.; Syms, R.R.A. NEMS by Sidewall Transfer Lithography. *J. Microelectromech. Syst.* **2014**, *23*, 1366–1373. [CrossRef]

40. Hållstedt, J.; Hellstrom, P.-E.; Radamson, H. Sidewall transfer lithography for reliable fabrication of nanowires and deca-nanometer MOSFETs. *Thin Solid Films* **2008**, *517*, 117–120. [CrossRef]

41. Sainiemi, L.; Franssila, S. Mask material effects in cryogenic deep reactive ion etching. *J. Vac. Sci. Technol. B Microelectron. Nanometer. Struct.* **2007**, *25*, 801. [CrossRef]

42. Moulder, J.F. *Hand Book of X-ray Photoelectron Spectroscopy*; Perkin Elmer Corporation: Eden Prairie, MN, USA, 1992; pp. 182–187.

43. Ghiraldelli, E.; Pelosi, C.; Gombia, E.; Chiavarotti, G.; Vanzetti, L. ALD growth, thermal treatments and characterisation of Al_2O_3 layers. *Thin Solid Films* **2008**, *517*, 434–436. [CrossRef]

44. Böse, O.; Kemnitz, E.; Lippitz, A.; Unger, W.E.S. C 1s and Au 4f7/2 referenced XPS binding energy data obtained with different aluminium oxides,-hydroxides and-fluorides. *Fresenius' J. Anal. Chem.* **1997**, *358*, 175–179.

45. Gehr, R.J.; Boyd, R.W. Optical Properties of Nanostructured Optical Materials. *Chem. Mater.* **1996**, *8*, 1807–1819. [CrossRef]

 © 2020 by the authors. Licensee MDPI, Basel, Switzerland. This article is an open access article distributed under the terms and conditions of the Creative Commons Attribution (CC BY) license (http://creativecommons.org/licenses/by/4.0/).

micromachines

Article

Metal-Assisted Chemical Etching and Electroless Deposition for Fabrication of Hard X-ray Pd/Si Zone Plates

Rabia Akan [1,*], Thomas Frisk [1], Fabian Lundberg [1], Hanna Ohlin [1], Ulf Johansson [2], Kenan Li [3], Anne Sakdinawat [3] and Ulrich Vogt [1]

[1]   KTH Royal Institute of Technology, Department of Applied Physics, Biomedical and X-ray Physics, Albanova University Center, 106 91 Stockholm, Sweden; tfrisk@kth.se (T.F.); fabianlu@kth.se (F.L.); hohli@kth.se (H.O.); ulrich.vogt@biox.kth.se (U.V.)
[2]   MAX IV Laboratory, Lund University, 22 100 Lund, Sweden; ulf.johansson@maxiv.lu.se
[3]   SLAC National Accelerator Laboratory, 2575 Sand Hill Road, Menlo Park, CA 94025, USA; kenan@stanford.edu (K.L.); annesak@stanford.edu (A.S.)
*   Correspondence: rabiaa@kth.se

Received: 28 February 2020; Accepted: 10 March 2020; Published: 13 March 2020

Abstract: Zone plates are diffractive optics commonly used in X-ray microscopes. Here, we present a wet-chemical approach for fabricating high aspect ratio Pd/Si zone plate optics aimed at the hard X-ray regime. A Si zone plate mold is fabricated via metal-assisted chemical etching (MACE) and further metalized with Pd via electroless deposition (ELD). MACE results in vertical Si zones with high aspect ratios. The observed MACE rate with our zone plate design is 700 nm/min. The ELD metallization yields a Pd density of 10.7 g/cm^3, a value slightly lower than the theoretical density of 12 g/cm^3. Fabricated zone plates have a grid design, 1:1 line-to-space-ratio, 30 nm outermost zone width, and an aspect ratio of 30:1. At 9 keV X-ray energy, the zone plate device shows a first order diffraction efficiency of 1.9%, measured at the MAX IV NanoMAX beamline. With this work, the possibility is opened to fabricate X-ray zone plates with low-cost etching and metallization methods.

Keywords: X-ray diffractive optics; zone plate; high aspect ratio nanostructures; metal-assisted chemical etching; electroless deposition

---

1. Introduction

X-ray microscopy is a powerful scientific tool used for the study of different samples in a variety of disciplines [1,2]. Especially in the hard X-ray regime (multi-keV photon energies), X-ray microscopy offers the possibility to study thick samples with nanoscale resolution. Diffractive zone plate optics are commonly used as focusing and imaging optics in these X-ray microscopes. Zone plates are circular, dense grating structures with radially decreasing zone widths [3,4]. The outermost zone width defines the zone plate resolution, whereas the zone thickness defines the zone plate diffraction efficiency. As an example, to focus X-rays with 9 keV photon energy with maximum efficiency performance, a zone plate made from Pd would require a thickness of 2.5 μm. The required high aspect ratios at very small zone widths make hard X-ray zone plates challenging to fabricate [5,6]. The performance of hard X-ray microscopes is therefore often limited by the fabrication quality of the focusing optics.

Two main high aspect ratio zone plate fabrication routes have been presented in the literature. The first is direct electron-beam patterning either of an organic or inorganic resist mold. The mold is then filled with a metal by electrodeposition [7,8] or covered with a metal layer by atomic layer deposition [9,10]. The second method transfers the zone plate pattern into a metal layer by deep reactive ion etching [11,12]. However, in both cases, the achievable aspect ratios are limited to about 20:1. Attempts have been made to overcome this limit by stacking of multiple zone plates, either mechanical [13,14] or via multi-level e-beam exposures [15,16], which is very challenging.

Metal-assisted chemical etching (MACE) has recently been used as an alternative fabrication method to transfer a zone plate pattern into a Si substrate with the advantage of vertical etching and the possibility to reach high aspect ratios [17–20]. MACE is a wet-chemical process where a noble metal layer, such as a Au zone plate pattern, is transferred into a Si substrate by an etching solution consisting of a strong oxidizing agent and HF [21–23]. The Si is predominantly oxidized where it is in contact with the noble metal layer, which catalyzes the oxidation process. The HF subsequently dissolves the formed SiO_2 [24,25].

Due to its low X-ray diffraction efficiency in the hard X-ray regime, the MACE processed Si zone plates are not suitable to be used directly as optical devices. Instead, they can be used as molds for high-Z materials, such as Pt, deposited with atomic layer deposition (ALD) [17,18]. Unfortunately, ALD is a rather slow and expensive process, and the homogeneous filling of deep and small trenches is challenging. Electrodeposition (ED) is an alternative wet-metallization process that is often used with polymer resist-based zone plate molds on conductive substrates [26]. ED is, however, challenging for metallizing MACE Si molds due to contacting difficulties to the Au layer at the bottom of the zone plate pattern. To overcome the challenge of contacting, the autocatalytic process of electroless deposition (ELD) is an alternative method to metalize the Si-based zone plates. In ELD, a metal complex is selectively reduced at a conductive surface [27,28]. The deposited metal now serves as a catalyst in the deposition reaction, ensuring continuous buildup of the metal.

In this work, we describe the fabrication process of hard X-ray Pd/Si zone plates using a simple and low-cost wet-chemical approach. MACE is used to transfer a zone plate pattern into a Si substrate, and ELD is used to metalize the Si zone plate mold with Pd. The Au zone plate pattern serves as a catalyst in both processes, under which the Si is selectively etched and on which the Pd is selectively deposited. Here, Pd is chosen as a high-Z material owing to an acceptable diffraction efficiency in the hard X-ray regime and the commercial availability of a well-formulated, stable ELD bath.

2. Materials and Methods

2.1. Materials

p-type, boron-doped Si (100) wafers with 1–5 Ωcm resistivity were purchased from Si-Mat. CSAR 62, amyl acetate, and dimethyl succinate were purchased from Allresist. Hydrofluoric acid (HF, 40%) and acetone were procured from Merck. Hydrogen peroxide (H_2O_2, 31%) and isopropanol were from D-BASF. The PD-Tech PC electroless Pd system was purchased from Atotech. Ethanol and *n*-pentane were from VWR.

2.2. Zone Plate Patterning Using Electron Beam Lithography (EBL)

The overall zone plate fabrication schematic is presented in Figure 1. The Si-wafers were cut into 1.5×1.5 cm^2 substrates and pre-cleaned under sonication in acetone followed by isopropanol for 5 min each. Thereafter, the substrates were cleaned in oxygen plasma for 5 min in an Oxford Instruments PlasmaLab 80 Plus RIE/ICP system. Seventy nanometer CSAR 62 EBL resist was spin-coated on the substrates and baked on a 180 °C hot plate for 60 s. The zone plate patterning was performed using a 50 kV Raith Voyager EBL system. The written zone plate pattern had a diameter of 150 µm, an outermost zone width of 30 nm, and a line-to-space ratio of 1:1. The exposed CSAR 62 was developed in amyl acetate for 60 s at room temperature. Further, the substrates were rinsed in isopropanol and n-pentane for 10 s and 15 s, respectively, and descummed in an oxygen plasma for 13 s for removal of exposed resist residues. Using an in-house Eurovac/Thermionics electron beam evaporation deposition system, a 2 nm adhesive Ti layer followed by a 10 nm Au layer were evaporated at a rate of 1 Å/s. The resist lift-off was performed in dimethyl succinate under sonication for 10 min, and the substrates were thereafter rinsed in isopropanol and deionized (DI) H_2O. The resulting substrates with Au zone plate patterns were dried under nitrogen gas.

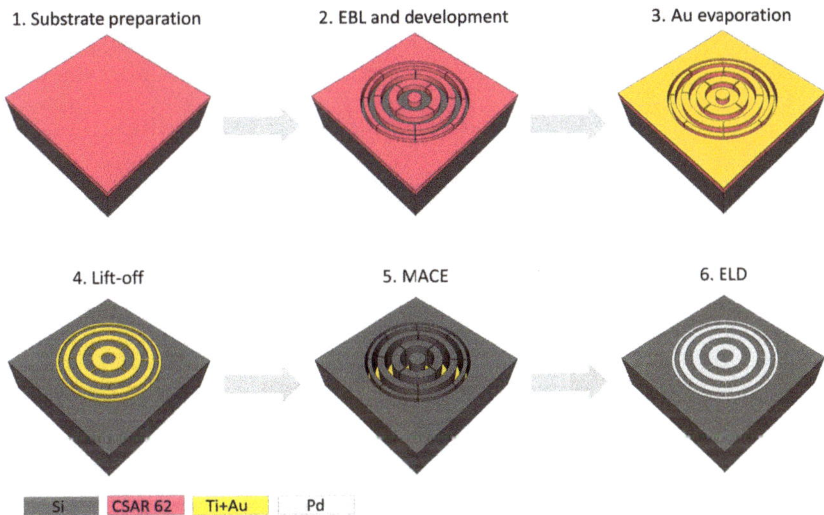

Figure 1. Schematic of the zone plate fabrication process.

2.3. MACE Processing of Si Zone Plate Molds

The Au patterned substrates were cleaned in oxygen plasma for 3 min right before MACE. The MACE zone plate processing was performed as reported previously [20]. Briefly, a 15 mL etching solution consisting of 0.68 M H_2O_2, 4.7 M HF, and 51 M DI H_2O was prepared in a polytetrafluoroethylene bath. The clean substrates were immersed in the etching solution, and the MACE process was performed at room temperature for 75 s under light protection. The substrates were rinsed in copious amounts of DI H_2O, transferred to ethanol, and dried in a Leica EM CPD300 critical point dryer.

2.4. Pd ELD Metallization of Zone Plates

The substrates were cleaned prior to ELD metallization in oxygen plasma for 3 min. This plasma oxidation step was found necessary for preventing Pd deposition on other sites than the catalytic Au layer at the bottom of the zone plate structures. A 100 mL plating solution consisting of 75 mL DI H_2O, 15 mL PD-Tech PC Reduction Solution, and 10 mL PD-Tech PC Plus Make-Up Solution was prepared as specified by Atotech and heated to 40 °C. After 30 min of temperature stabilization, the Si substrates were vertically immersed in the plating solution, and the ELD proceeded for 30 min under stirring. After completion of the ELD, the substrates were rinsed in DI H_2O and dried under nitrogen gas.

2.5. Characterization

Micrographs were obtained and cross-sections prepared with an FEI NOVA 200 dual-beam scanning electron microscopy (SEM) and focused ion beam (FIB) system.

The static contact angle measurements were performed using a Biolin Scientific Theta Lite Optical Tensiometer on 10 nm Au-coated Si substrates before and after a 3 min oxygen plasma treatment. At room temperature, a 5 μL DI H_2O droplet was released on the substrate, and the contact angles on both sides of the droplet were continuously recorded. The presented data were the average value of the left and right contact angles over a 4 s contact time.

The Pd density was determined gravimetrically using a Sartorius analytical scale and a KLA Tencor P-15 surface profiler by weighing and measuring the thicknesses of several Pd films plated on 10 nm Au-coated Si plating bases.

The zone plate diffraction efficiency was quantified at the NanoMAX beamline at the MAX IV synchrotron radiation facility [29]. Several zone plates were illuminated by a coherent beam with a photon energy of 9 keV. A 500 µm aperture with a 25 µm-wide and 25 µm-thick tungsten central stop was placed before a zone plate, and a 10 µm order sorting aperture (OSA) was placed slightly upstream of the zone plate focal plane. The first order diffraction was recorded by a Crycam X-ray camera from Crytur. For the calculation of the first order diffraction efficiency, several images with the first order zone plate cone and the empty beam without the zone plate and OSA (but transmitted through the Si substrate) were recorded.

3. Results and Discussion

3.1. Zone Plate Fabrication

As reported previously, a Au zone plate catalyst design with interconnects between the zones plate rings is essential to get a controlled, vertical etching during MACE at ambient processing conditions (Figure 2a) [20]. While zone verticality is ensured with this design, too thick zones will tilt due to mechanical instabilities. This is especially visible at the outermost parts of the zone plate where the zone widths are the smallest. Therefore, the deepest zone thickness was chosen as 900 nm in this study, giving an aspect ratio of 30:1 (Figure 2b). For 150 µm zone plates with a 10 nm Au catalyst layer, the observed MACE rate was 700 nm/min.

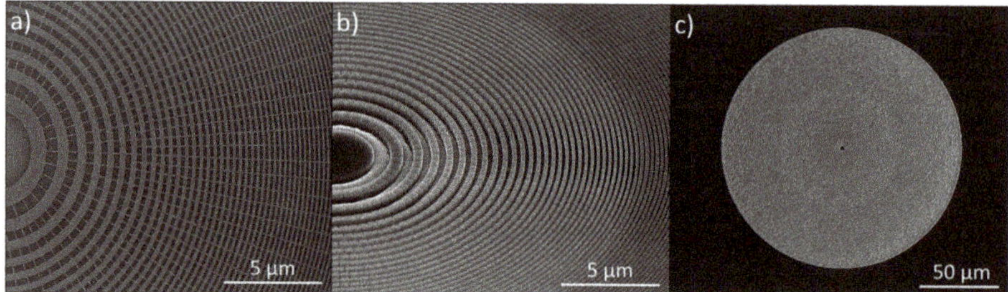

Figure 2. SEM micrographs of (**a**) Au patterned Si substrate (top-view), (**b**) MACE processed Si zone plate mold (52°-tilt view), and (**c**) ELD metalized Pd/Si zone plate showing selective plating in the zone plate area only (top-view).

Measurements revealed a decrease of the contact angle for the oxygen plasma-treated Au films, indicating a more hydrophilic surface character and easier wetting. The measured contact angles before and after the oxygen plasma treatment were 89° and 29°, respectively. In addition, no plating was observed on sites other than on the Au at the bottom of the Si zone plate mold (Figure 2c). This suggested that the short plasma treatment was enough to grow a passivating oxide layer on the Si substrate. Thirty nanometer wide zones with thicknesses of 3 µm were evenly plated, and no limitations were observed with the Pd ELD process. These results are however not shown here due the stability issues of the Si nanostructures, resulting in tilted zones.

The decreasing zone widths resulted in a variation of zone thickness over the zone radius. The inner zones were wider than the outer zones and consequently relatively thin compared to the outer zones. In order to fill the zones fully with Pd via ELD, over-plating was unavoidable. The variation of zone thickness over the zone plate radius is presented in Figure 3a. The zone thickness variation was largest from the zone plate center to 20 µm outwards. However, this variation was not a big concern for the final zone plate performance since, normally, a central stop is used. The zone thickness was uniform from 20 µm to the outermost part of the zone plate, with a standard deviation of about ± 30 nm. Cross-section SEM micrographs are presented in Figure 3b for the inner, middle, and outer parts of a typical Pd/Si zone plate. The tendency of non-vertical zones is visible for the 30 nm wide zones in the outer part, where the Si structures are slightly slanted.

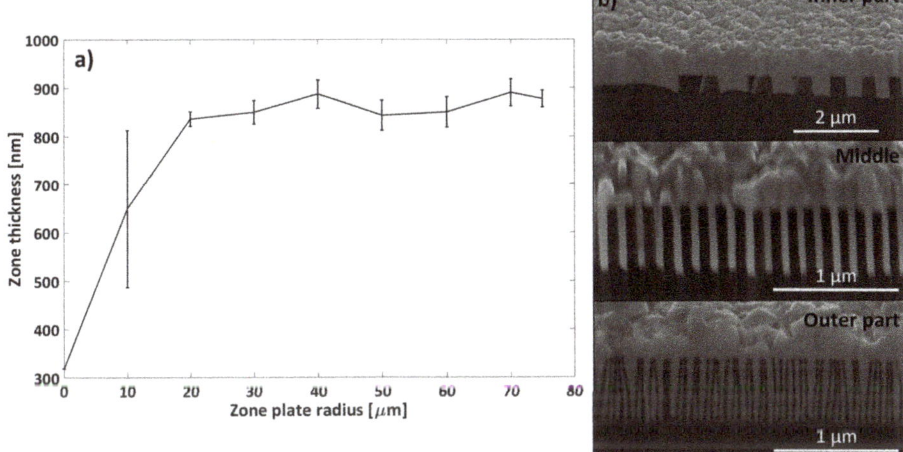

Figure 3. (**a**) Zone thickness profile over the zone plate radius based on image analysis of SEM cross-section micrographs. The error bars represent the zone thickness variation of the surrounding zones at each point. (**b**) SEM micrographs of the cross-sections of the inner, middle, and outer parts of a typical Pd/Si zone plate (52°-tilt view).

3.2. Pd Density

A pure Si zone plate would require extreme thicknesses to serve directly as a zone plate device. At a 900 nm zone thickness at 9 keV, a first order diffraction efficiency of only 0.6% would be expected (Figure 4a). Oppositely, a pure Pd zone plate has a theoretical efficiency of 10% at the same thickness. A combined Pd/Si zone plate, as in our work, would show a shift in focusing performance so that thicker zones would be required to reach the maximum efficiency. The 30% maximum diffraction efficiency would be expected at a 3 μm thickness for a combined Pd/Si zone plate given a tabulated Pd density of 12 g/cm^3. We kept the Si walls in our zone plates after filling with Pd due to the mechanical support that they provided. For a 900 nm thick Pd/Si zone plate with tabulated Pd density, an efficiency of 6.4% was expected.

The density of the ELD plated Pd was gravimetrically determined to be 10.7 ± 0.4 g/cm^3. With lower Pd density, the expected zone plate efficiency also decreased (Figure 4b). It should be noted that a lower density than the tabulated value was expected for ELD plated Pd. Pores, voids, and impurities in the Pd deposit were common reasons for this. The inclusion of relatively light-weight elements in the ELD bath formulation, such as brighteners, would decrease the deposit density [30]. The expected efficiencies for 900 nm Pd/Si zone plates with Pd densities of 10 g/cm^3 and 11 g/cm^3 were 4.0% and 5.1%, respectively.

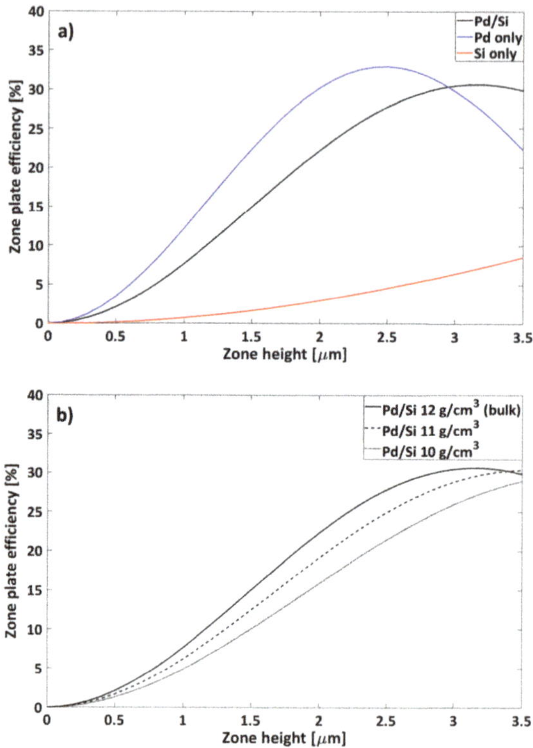

Figure 4. Zone plate efficiency at 9 keV as a function of zone height for (**a**) Pd/Si, Pd only, and Si only zone plates (bulk densities) and (**b**) Pd/Si zone plates with different Pd densities and bulk Si density. Calculations were done in MATLAB with GD-Calc.

3.3. Focusing Performance

Figure 5 shows an image of the first order diffraction cone and illustrates the local zone plate efficiency. The best measured total zone plate efficiency of our 900 nm thick devices at 9 keV was 1.9%. The main factor for the lower zone plate efficiency compared to theoretical values could be explained by a decrease of the local efficiency towards larger radii. The local efficiency was even to about 2/3 of the radius, and then gradually dropped outwards. This was believed to be due to the more tilted zones in the outer part (cf. Figure 3b). Moreover, the Au catalyst had a grid-like design with interconnects between the rings (Figure 2a). While this grid design was necessary to ensure zone verticality during the MACE process, the interconnects reduced the effective zone plate area. Lastly, the Pd overplating had a minor effect on the efficiency due to some absorption of the incoming photons.

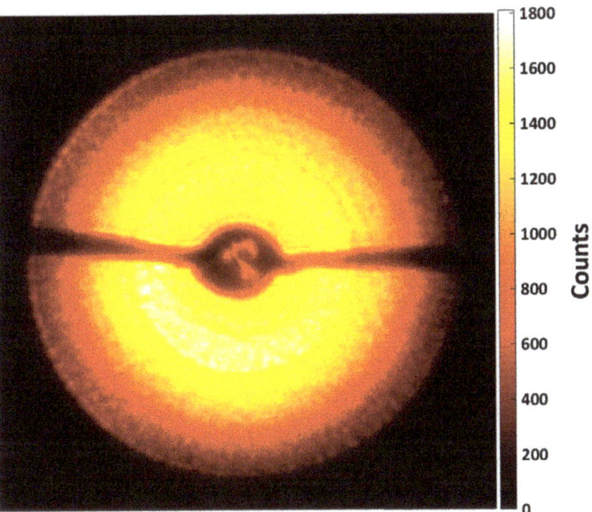

Figure 5. Raw image of the first order diffraction cone showing a map of the local zone plate efficiency.

4. Conclusions

In this paper, we presented a wet-chemical route to fabricate high aspect ratio Pd/Si zone plates aimed for the hard X-ray regime. MACE was used to fabricate a Si zone plate mold, and ELD was used to metalize the mold with Pd. We demonstrated and characterized the zone plates using this fabrication route. The optics device had a grid design with a 30 nm outermost zone width and a 900 nm zone thickness, thus an aspect ratio of 30:1. The lower zone plate efficiency of 1.9%, compared to theory, was mainly due to the tilt of the outermost zones, the loss of effective zone plate area by the grid design, and some photon absorption due to Pd overplating.

The MACE parameters used in this study were optimized for processing at ambient conditions with the grid zone plate design [20]. Vertically etched zones were obtained at the presented process conditions; however, some tilting was observed in the outermost parts. This was due to the mechanical instabilities of the free-standing Si structures with the grid design, and the aspect ratio was therefore limited to 30:1. To reach larger aspect ratios, a different zone plate design with interconnected Si structures should be adapted, as reported by Chang and Sakdinawat [17]. Combined with Pd electroless deposition, much higher efficiencies should be possible.

Author Contributions: Conceptualization, R.A., A.S., and U.V.; methodology, R.A. and U.V.; investigation, R.A., T.F., H.O., F.L., U.J., K.L., and U.V.; visualization, R.A.; writing, original draft preparation, R.A. and U.V. All authors have read and agreed to the published version of the manuscript.

Funding: This research was funded by the Swedish Research Council Grant Number 2018-04273 and the Olle Engkvists stiftelse Grant Number 196-0068.

Acknowledgments: We would like to thank Muhammet S. Toprak (KTH) for useful discussions; Alexander Björling and Sebastian Kalbfleisch for their help with the experiments at the MAX IV NanoMAX beamline; and Hazal Batili (KTH) for her help with the MACE experiments. We acknowledge the support from SLAC National Accelerator Laboratory. Parts of this work were performed at the Albanova Nanofabrication Facility (ANF) and the Stanford Nano Shared Facilities (SNSF).

Conflicts of Interest: The authors declare no conflict of interest.

Abbreviations

The following abbreviations are used in this manuscript:

MACE   Metal-assisted chemical etching
ALD   Atomic layer deposition
ED   Electrodeposition
ELD   Electroless deposition
EBL   Electron beam lithography
DI   Deionized
SEM   Scanning electron microscopy
FIB   Focused ion beam
OSA   Order sorting aperture

References

1. Sakdinawat, A.; Attwood, D. Nanoscale X-ray imaging. *Nat. Photonics* **2010**, *4*, 840–848. [CrossRef]
2. Ice, G.E.; Budai, J.D.; Pang, J.W.L. The Race to X-ray Microbeam and Nanobeam Science. *Science* **2011**, *334*, 1234–1239. [CrossRef] [PubMed]
3. Soret, J.L. Ueber die durch Kreisgitter erzeugten Diffractionsphänomene. *Ann. Der Phys.* **1875**, *232*, 99–113. [CrossRef]
4. Attwood, D.; Sakdinawat, A. *X-rays and Extreme Ultraviolet Radiation*; Cambridge University Press: Cambridge, UK, 2016; doi:10.1017/CBO9781107477629. [CrossRef]
5. Wu, S.R.; Hwu, Y.; Margaritondo, G. Hard-X-ray Zone Plates: Recent progress. *Materials* **2012**, *5*, 1752–1773. [CrossRef]
6. Lider, V.V. Zone Plates for X-Ray Focusing (Review). *J. Surf. Investig.* **2017**, *11*, 1113–1127. [CrossRef]
7. Gorelick, S.; Vila-Comamala, J.; Guzenko, V.; Mokso, R.; Stampanoni, M.; David, C. Direct e-beam writing of high aspect ratio nanostructures in PMMA: A tool for diffractive X-ray optics fabrication. *Microelectron. Eng.* **2010**, *87*, 1052–1056. [CrossRef]
8. Zhu, J.; Chen, Y.; Xie, S.; Zhang, L.; Wang, C.; Tai, R. Nanofabrication of 30 nm Au zone plates by e-beam lithography and pulse voltage electroplating for soft X-ray imaging. *Microelectron. Eng.* **2020**, *225*, 111254. [CrossRef]
9. Vila-Comamala, J.; Jefimovs, K.; Raabe, J.; Pilvi, T.; Fink, R.H.; Senoner, M.; Maassdorf, A.; Ritala, M.; David, C. Advanced thin film technology for ultrahigh resolution X-ray microscopy. *Ultramicroscopy* **2009**, *109*, 1360–1364. [CrossRef]
10. Vila-Comamala, J.; Gorelick, S.; Guzenko, V.; Färm, E.; Ritala, M.; David, C. Dense high aspect ratio hydrogen silsesquioxane nanostructures by 100 keV electron beam lithography. *Nanotechnology* **2010**, *21*, 285305. [CrossRef]
11. Uhlén, F.; Lindqvist, S.; Nilsson, D.; Reinspach, J.; Vogt, U.; Hertz, H.M.; Holmberg, A.; Barrett, R. New diamond nanofabrication process for hard X-ray zone plates. *J. Vac. Sci. Technol. B Nanotechnol. Microelectron. Mater. Process. Meas. Phenom.* **2011**, *29*, 06FG03. [CrossRef]
12. Parfeniukas, K.; Rahomäki, J.; Giakoumidis, S.; Seiboth, F.; Wittwer, F.; Schroer, C.G.; Vogt, U. Improved tungsten nanofabrication for hard X-ray zone plates. *Microelectron. Eng.* **2016**, *152*, 6–9. [CrossRef]
13. Feng, Y.; Feser, M.; Lyon, A.; Rishton, S.; Zeng, X.; Chen, S.; Sassolini, S.; Yun, W. Nanofabrication of high aspect ratio 24 nm X-ray zone plates for X-ray imaging applications. *J. Vac. Sci. Technol. B: Microelectron. Nanometer Struct.* **2007**, *25*, 2004–2007. [CrossRef]
14. Gleber, S.C.; Wojcik, M.; Liu, J.; Roehrig, C.; Cummings, M.; Vila-Comamala, J.; Li, K.; Lai, B.; Shu, D.; Vogt, S. Fresnel zone plate stacking in the intermediate field for high efficiency focusing in the hard X-ray regime. *Opt. Express* **2014**, *22*, 28142. [CrossRef] [PubMed]
15. Werner, S.; Rehbein, S.; Guttmann, P.; Heim, S.; Schneider, G. Towards high diffraction efficiency zone plates for X-ray microscopy. *Microelectron. Eng.* **2010**, *87*, 1557–1560. [CrossRef]
16. Mohacsi, I.; Vartiainen, I.; Rösner, B.; Guizar-Sicairos, M.; Guzenko, V.A.; McNulty, I.; Winarski, R.; Holt, M.V.; David, C. Interlaced zone plate optics for hard X-ray imaging in the 10 nm range. *Sci. Rep.* **2017**, *7*, 1–10. [CrossRef]

17. Chang, C.; Sakdinawat, A. Ultra-high aspect ratio high-resolution nanofabrication for hard X-ray diffractive optics. *Nat. Commun.* **2014**, *5*, 1–7. [CrossRef]
18. Li, K.; Wojcik, M.J.; Divan, R.; Ocola, L.E.; Shi, B.; Rosenmann, D.; Jacobsen, C. Fabrication of hard X-ray zone plates with high aspect ratio using metal-assisted chemical etching. *J. Vac. Sci. Technol. B Nanotechnol. Microelectron. Mater. Process. Meas. Phenom.* **2017**, *35*, 06G901. [CrossRef]
19. Tiberio, R.C.; Rooks, M.J.; Chang, C.; Knollenberg, C.F.; Dobisz, E.A.; Sakdinawat, A. Vertical directionality-controlled metal-assisted chemical etching for ultrahigh aspect ratio nanoscale structures. *J. Vac. Sci. Technol. B Nanotechnol. Microelectron. Mater. Process. Meas. Phenom.* **2014**, *32*, 06FI01. [CrossRef]
20. Akan, R.; Parfeniukas, K.; Vogt, C.; Toprak, M.S.; Vogt, U. Reaction control of metal-assisted chemical etching for silicon-based zone plate nanostructures. *RSC Adv.* **2018**, *8*, 12628–12634. [CrossRef]
21. Huang, Z.; Geyer, N.; Werner, P.; De Boor, J.; Gösele, U. Metal-assisted chemical etching of silicon: A review. *Adv. Mater.* **2011**, *23*, 285–308. [CrossRef]
22. Li, X. Metal assisted chemical etching for high aspect ratio nanostructures: A review of characteristics and applications in photovoltaics. *Curr. Opin. Solid State Mater. Sci.* **2012**, *16*, 71–81. [CrossRef]
23. Han, H.; Huang, Z.; Lee, W. Metal-assisted chemical etching of silicon and nanotechnology applications. *Nano Today* **2014**, *9*, 271–304. [CrossRef]
24. Chartier, C.; Bastide, S.; Lévy-Clément, C. Metal-assisted chemical etching of silicon in HF-H2O2. *Electrochim. Acta* **2008**, *53*, 5509–5516. [CrossRef]
25. Peng, K.; Lu, A.; Zhang, R.; Lee, S.T. Motility of metal nanoparticles in silicon and induced anisotropic silicon etching. *Adv. Funct. Mater.* **2008**, *18*, 3026–3035. [CrossRef]
26. Bicelli, L.P.; Bozzini, B.; Mele, C.; D'Urzo, L. A review of nanostructural aspects of metal electrodeposition. *Int. J. Electrochem. Sci.* **2008**, *3*, 356–408.
27. den Exter, M.J. *The Use of Electroless Plating as a Deposition Technology in the Fabrication of Palladium-Based Membranes*; Woodhead Publishing Limited: Cambridge, UK, 2014; pp. 43–67. [CrossRef]
28. Deckert, C.A. Electroless copper plating. A review: Part I. *Plat. Surf. Finish.* **1995**, *82*, 58–64.
29. Johansson, U.; Vogt, U.; Mikkelsen, A. NanoMAX: A hard X-ray nanoprobe beamline at MAX IV. *X-ray Nanoimaging Instrum. Methods* **2013**, *8851*, 88510L. [CrossRef]
30. Wilkinson, P. Understanding gold plating. *Gold Bull.* **1986**, *19*, 75–81. [CrossRef]

© 2020 by the authors. Licensee MDPI, Basel, Switzerland. This article is an open access article distributed under the terms and conditions of the Creative Commons Attribution (CC BY) license (http://creativecommons.org/licenses/by/4.0/).

 *micromachines*

Article

Investigation of the Pd Nanoparticles-Assisted Chemical Etching of Silicon for Ethanol Solution Electrooxidation

Olga Volovlikova [1,*], Gennady Silakov [1], Sergey Gavrilov [1], Tomasz Maniecki [2] and Alexander Dudin [3]

1   Institute of Advanced Materials and Technologies, National Research University of Electronic Technology (MIET), Moscow 124498, Russia; mr.komrad-13@ya.ru (G.S.); pcfme@miee.ru (S.G.)
2   Institute of General and Ecological Chemistry, Lodz University of Technology, Zeromskiego 116, 90–924 Lodz, Poland; tmanieck@p.lodz.pl
3   Institute of Nanotechnology of Microelectronics of the Russian Academy of Sciences (INME RAS), Moscow115487, Russia; dudin_aa@icloud.com
*   Correspondence: 5ilova87@gmail.com; Tel.: +7-903-292-1507

Received: 7 November 2019; Accepted: 10 December 2019; Published: 12 December 2019

Abstract: The formation of porous silicon by Pd nanoparticles-assisted chemical etching of single-crystal Si with resistivity $\rho = 0.01$ $\Omega{\cdot}$cm at 25 °C, 50 °C and 75 °C in $HF/H_2O_2/H_2O$ solution was studied. Porous layers of silicon were studied by optical and scanning electron microscopy, and gravimetric analysis. It is shown that por-Si, formed by Pd nanoparticles-assisted chemical etching, has the property of ethanol electrooxidation. The chromatographic analysis of ethanol electrooxidation products on por-Si/Pd shows that the main products are CO_2, CH_4, H_2, CO, O_2, acetaldehyde $(CHO)^+$, methanol and water vapor. The mass activity of the por-Si/Pd system was investigated by measuring the short-circuit current in ethanol solutions. The influence of the thickness of porous silicon and wafer on the mass activity and the charge measured during ethanol electrooxidation was established. Additionally, the mechanism of charge transport during ethanol electrooxidation was established.

Keywords: porous silicon; Pd nanoparticles-assisted chemical etching; etching rate; ethanol electrooxidation

---

1. Introduction

Technological innovation leads to an increase in energy and natural resources consumption, in particular, natural gas. Continuous consumption of non-renewable resources leads to their exhaustion. There is a need to develop and make use of alternative energy sources, based on environmentally-friendly technologies, because of natural resources limitations. Development of new alternative energy sources will allow us to receive scientific and technical results, with technologies which provide transition to resource-saving energy. One of the prospective directions in the resource-saving power area is the fuel element and portable electrochemical energy generators. They have many advantages: portability, high efficiency, small level of harmful emissions and quietness [1]. Nowadays, an interesting and prospective direction in the resource-saving power area is ethanol as fuel and a power source for the electric current generators [2].

Transformation to electric power is due to direct ethanol oxidation in the cell. It allows the simplification of the fuel supply system because of high specific energy of liquid alcohols, providing a short circuit in an environmentally-friendly cycle of transformation of energy in the natural scale due to a number of alcohols. The ethanol can be produced in biosystems in almost unlimited volumes [3]. The oxidation product of ethanol is CH_4.

Recent studies for alternative fuels indicate a growing interest in the development of small fuel cells and energy generators based on porous silicon due to a few advantages: high specific surface; strong chemical loading ability of the surface; the possibility of changing the surface morphology of the porous layers at the nano- and micro levels; simplicity and low cost of manufacture and compatibility with silicon integrated technology [4]. The plasma chemical etching for porous silicon formation is widely used. This method is characterised by a high complexity of hardware provided [5]. Chemical etching in solutions of alkalis or acids is a cheaper method for porous Si formation. Porous Si is usually formed by anode etching in HF solutions [6–8]. However, this method means individual treatment of wafers.

In recent years, special development was received by the chemical etching induced by noble metals (Ag, Au, Pt, and Pd) [9–12]. This method is simple and enables carrying out group treatment of wafers that reduces the price of the technology of porous silicon formation. Besides, this method allows the production of porous silicon with a wide range of geometrical sizes by using a form and type of a metal mask. The thickness of the porous silicon is defined by the etching duration, electrolyte composition, and the metal amount. The MACE (metal-assisted chemical etching) method enables the production of a noble metal/porous silicon structure that combines the functions of both the anode and cathode of the generator's active element and fuel cell. This structure is called the Schottky junction (Schottky barrier). In work by Bin Zhu [13,14] a hydrogen fuel cell based on the Schottky barrier (metal, p-type semiconductor) was described.

Porous silicon functionalised with noble metals is of undoubted scientific and practical interest as an object for the production of new energy generators. Simplifying the design by switching to a single-layer functional structure will increase the productivity and reduce the cost of the finished device. Conducting studies of the electrocatalytic ethanol oxidation using the cathode/anode structure, based on Pd clusters in a porous layer, allows producing an environmentally-friendly and resource-saving energy sector. The efficiency of creating a structure is associated with a high specific surface area of the porous silicon, and size effects of a metal catalyst during the ethanol electrooxidation. It occurs on the local nano- and micro-regions of the anodes/cathodes inside of porous silicon. Understanding the electro-catalytic activities of the ethanol on the Pd/por-Si structure is very important for developing more active catalysts for the direct-ethanol generators. The purpose of the work is establishment of the influence of the porous layer thickness and porosity formed by Pd nanoparticles-assisted etching on the duration of gas evolution, and ethanol electrooxidation mass activity for different ethanol-based electrolytes.

2. Materials and Methods

Boron doped p-type silicon wafers with (1 0 0) orientation and resistivity of 0.01 $\Omega \cdot$cm were cleaned as follows: (i) dipping into H_2SO_4 (98%):H_2O_2 (30%) (in volume 1:1) solution, then (ii) into HF (40%):H_2O (in volume 1:4) solution to remove native oxide and finally, (iii) into pure ethanol. The cleaned wafers were cut into pieces 3×3 cm^2. The samples were placed in a Teflon cell. The samples were immersed into solution PdCl$_2$:HCl (0.5 g/L PdCl$_2$, 20 mL/L HCl) for 30 min at 25 °C for Pd film deposition. The por-Si formation was performed in solution HF (40%):H_2O_2 (30%):H_2O (25:10:4 in volume) at $T = 25, 50$ and 75 °C. Porosity was calculated by gravimetric analysis. Samples with area S were weighed before etching (m_1). Then non-polished side of the sample was covered by varnish and dried in the air. The etching duration was 30–120 min. Then, samples were cleaned in ethanol and dried at 65 °C. The varnish was removed and the samples were weighed (m_2). Por-Si was dissolved in a water solution of NaOH and weighed (m_3). Porosity was calculated by the equation:

$$P = \frac{m_1 - m_2}{m_1 - m_3} \times 100\%, \tag{1}$$

where m_1 and m_2 are the sample's masses before and after etching and m_3 is the sample mass after por-Si dissolution. Samples were etched in the same conditions because of the multi-sectional Teflon cell.

The duration of filling of the same gas volume (2.4 mL) during ethanol electrooxidation (EEO), with different C_2H_5OH concentrations, was fixed. The gas volume, as a result of the ethanol electrooxidation by 2.5 cm^2 samples, was measured using a cell (Figure 1). The essence of the method is the displacement of water by the gas, under a glass cell.

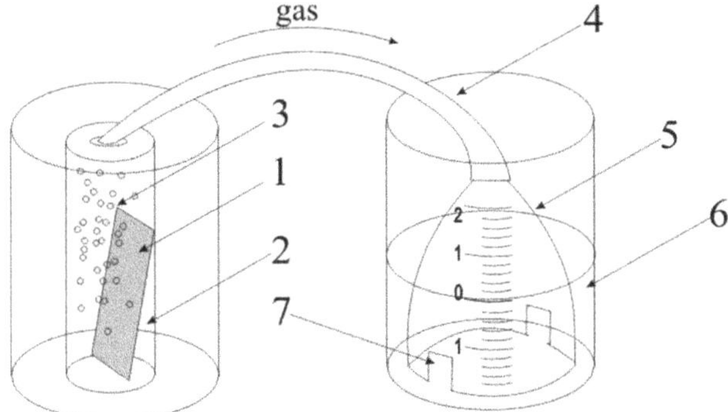

Figure 1. Schematic presentation of cell for gas volume measurement: (**1**) sample, (**2**) ethanol solution, (**3**) gas, (**4**) flexible tube, (**5**) cell, (**6**) glass cell, (**7**) holes.

The sample (1) was placed into a glass cell with ethanol solution (2). Gas evolution (3) occurs. Then, gas follows on a flexible tube (4) in cell (5) located in glass cell (6) with water (level 0). Then the gas displaces water from cell through the holes (7) and the level changed. The displaced water volume equals to gas volume evolved during a certain time $V_{water} = V_{gas}$.

The cell was filled multiple times to calculate the gas evolution rate v. It was calculated by the equation:

$$v = \frac{V_{gas}}{t} \tag{2}$$

where V_{gas} is the gas volume and t is the duration of the gas evolution.

Sample surface morphology was investigated by optical and scanning electron microscopy (Carl Zeiss Axiovert 40 MAT (Carl Zeiss Group, Oberkochen, Germany) and Helios NanoLab 650 (Thermo Fisher Scientific, Hillsboro, OR, USA)). Energy-dispersive X-ray (EDX) spectroscopy was performed on FEI Helios (FEI, Hillsboro, OR, USA) with an EDAX Octane Elite super EDS System (Octane Super, Mahwah, NJ, USA). The short-circuit current during etching in the galvanic cells and the mass activity as a function of EEO time were measured with a digital multimeter (UNI-T UT61C, UNI-T Group Ltd., Hong Kong). The value of the charge $Q_{Excess\ Carrier}$ was determined by numerical integration of current versus time, $Q = \int_0^t J dt$. The composition and conductivity of ethanol solutions for mass activity measurements are presented in Table 1. pH equals to 2 made by adding H_2SO_4 into ethanol solution.

Table 1. Solution properties.

Solution	Ethanol/Water (Volume Ratio)	pH	Conductivity
1	10/90	2	5.00 mS/cm
2	50/50	2	2.84 mS/cm
3	95/5	2	0.514 mS/cm

The gas composition was measured using quadrupole mass selective detector Hiden Analytical HPR-20 (Hiden Analytical, Warrington, UK) in the m/z range 1–100. EEO was performed in glass reactor at $T = 25\ °C$, $p = 1$ bar. To obtain information on types of chemical bonds presented in the porous layer, the samples were analyzed by infrared (IR) reflectance spectroscopy using Nicolet iS50 spectrometer (Thermo Fisher Scientific).

3. Results

Figure 2 shows scanning electron micrograph (SEM) images of Si (1 0 0) surface with palladium clusters deposited during immersion in $PdCl_2$ solution for 30 min. The dimension of separate Pd particles varied in the range of $20 < d < 50$ nm. The agglomerates from the Pd particles varied in the range of $0.1 < d < 2$ μm.

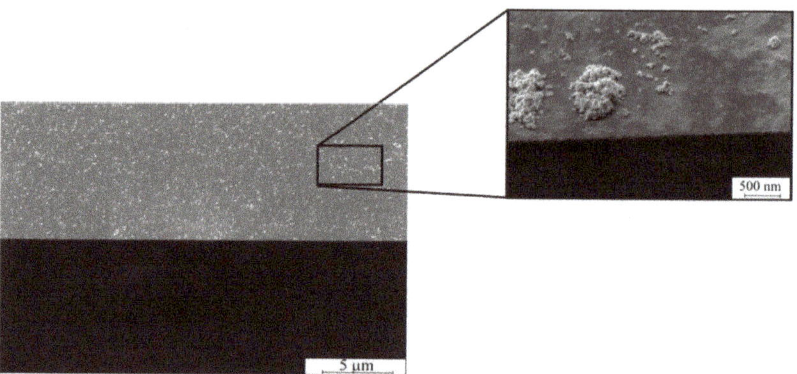

Figure 2. Scanning electron micrograph (SEM) image of Pd particles deposited on Si (1 0 0) wafer by immersing in $PdCl_2$:HCl solution for $t = 30$ min.

Figure 3 shows SEM images of the porous silicon surface after Pd nanoparticles-assisted chemical etching with different etching times.

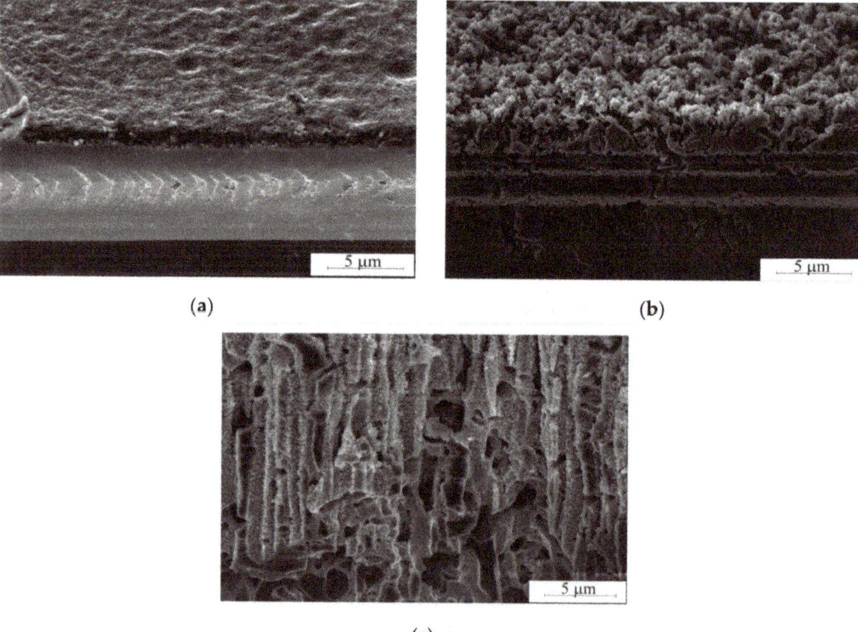

Figure 3. Scanning electron micrograph of the cross section porous silicon film etched for (**a**) 2 min, (**b**) 15 min and (**c**) 120 min at 25 °C.

Figure 3 illustrates porous silicon formation during different etching times. The minimum etching duration was 2 min because of 1.25 μm porous layer is observed. The maximum etching duration was 120 min. The porous layer breaks after 120 min etching. The porous silicon has perpendicular macropores with diameters from 1 to 3 μm and mesopores tightly penetrating the walls of the porous matrix. Pd particles dissolve silicon, gradually plunging inwards and forming a pore. Parts of the particles were deposited on the pore walls during the etching process. This contributed to the formation of mesopores.

Each pore is characterised by the size and shape of the metal particle. The appearance of the inner pore cavity is a consequence of the etching. If the particle is agglomerated, the pore walls are like a sponge of mesoporous silicon. As the agglomerate moves deeper into the pore, individual particles with a diameter of 80 ± 5 nm are deposited on the pore wall, contributing to its dissolution. This phenomenon is described in [12]. As particles are deposited on the walls, the agglomerate decreases in size, forming a dimple of a smaller diameter, which ultimately leads to the formation of a conical pore. In this case, the conical pore is due to the gradual dissolution of the walls (Figure 4a). Etching segments can determine the particle size. The upper part of the pore promotes its expansion with the solution penetrating the portion, uniformly dissolving the walls of the pore by photoelectrochemical dissolution. In the case of the separate metal particles, the pores will be formed vertically (Figure 4b). The pore diameter will be equal to the particle diameter.

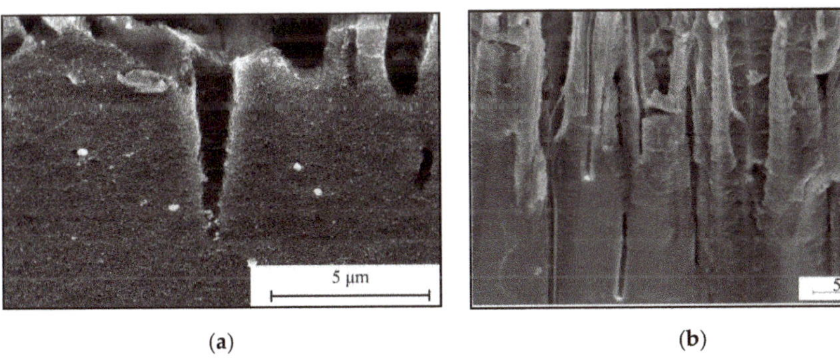

(a)   (b)

Figure 4. SEM image of cross section of sample after etching during 60 min: (**a**) the Pd particles are agglomerated, (**b**) the separate Pd particles.

It was found that Pd particles are present in the mesopores, which have access to a system por-Si/Pd. This system provides the ethanol electrooxidation (EEO). It can proceed according to one of a few schemes (Figure 5) [15]. The acetaldehyde, acetic acid (incomplete oxidation) and carbon dioxide (complete oxidation) are the products of electrooxidation of ethanol [16].

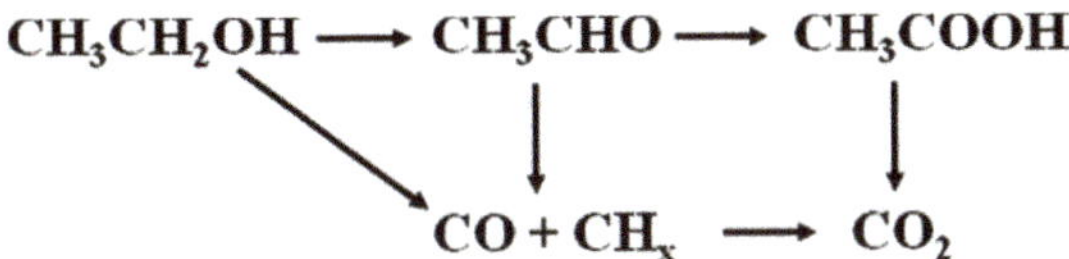

$$CH_3CH_2OH \longrightarrow CH_3CHO \longrightarrow CH_3COOH$$
$$CO + CH_x \longrightarrow CO_2$$

Figure 5. The scheme of the ethanol electrooxidation.

Figure 6a illustrates EEO. Intense gas evolution is observed from the functionalised metal-based silicon porous material immersed in ethanol solutions. Intense gas evolution gradually decreased. The gas evolution duration can be as long as several hours. The gas composition is shown in Figure 6b.

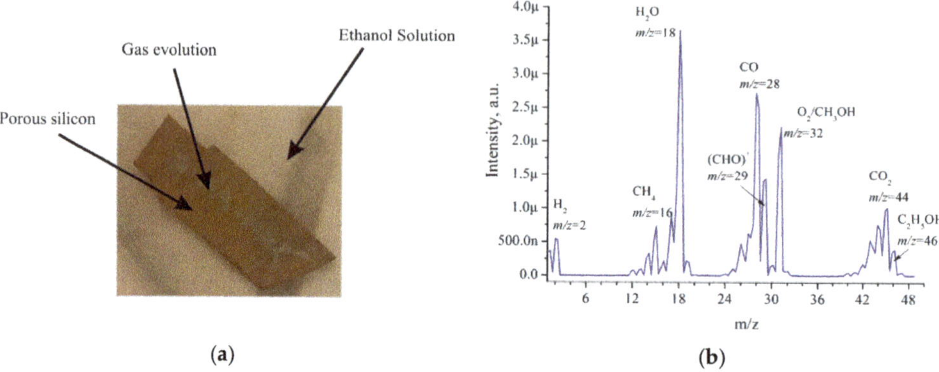

(**a**)   (**b**)

Figure 6. Electrooxidation (EEO): (**a**) sample photo during EEO, (**b**) the mass spectrum of a gas.

The mass spectrometry analysis of EEO products on por-Si/Pd shows that the main products are CO_2, CH_4, H_2, CO, O_2, acetaldehyde $(CHO)^+$ [17], methanol, ethanol and water vapor. The volumes of components relative to the total volume of EEO products are the following: H_2—5%, CH_4—7%, H_2O—33%, CO—25%, O_2—20%, CO_2—10%. Figure 7a,b shows the influence of the sample porosity, etching duration and temperature on the duration of gas evolution. In this work, intensive gas evolution can be visually detected without using additional devices.

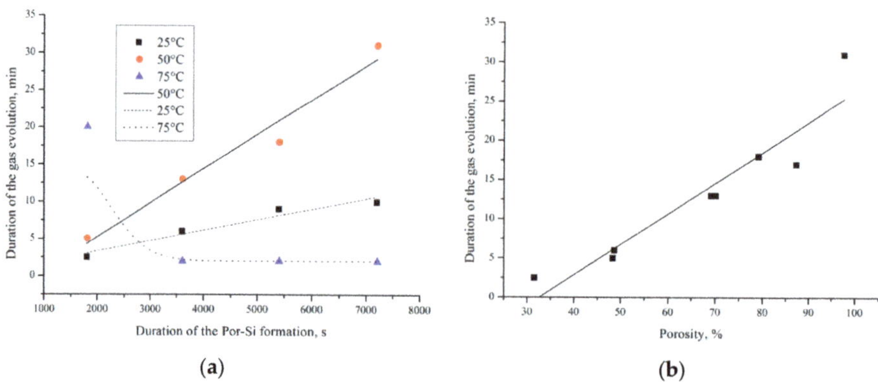

(**a**)   (**b**)

Figure 7. The dependence of the gas evolution on the (**a**) samples treatment duration, (**b**) por-Si porosity. Points are experimental results, lines are approximations.

It has been found that the maximum duration of intense gas evolution (32 min) is observed for samples formed at 50 °C for 120 min etching, and the minimum duration (2 min) is observed for samples formed at 75 °C for 60–120 min etching. Increase and reduction of the duration of gas evolution is caused by the increase and reduction of porous silicon thickness respectively. It was established that a linear increase in the thickness of the porous layer happened with an increase in the etching duration from 30 to 120 min for temperatures of 25 and 50 °C. The thickness of the porous layer was from 30 to 90 μm for 30–120 min and 25 °C etching; from 60 to 105 μm for 30–120 min and 50 °C etching; and from 100 to 45 μm for 30–120 min and 75 °C etching. It is due to porous layer dissolution. Reducing the porous silicon thickness leads to a decrease of the local metal/semiconductor and the EEO regions, respectively. The transfer of porous silicon at elevated processing temperatures is described in detail in [18]. The duration of the gas evolution is linearly dependent on the porosity of the layer. Therefore, the high porosity of the sample ensures access of the reactants to the surface of por-Si/Pd and removal of the reaction products.

Besides the duration of gas evolution, an important parameter for establishing the EEO mechanism is the gas evolution rate. It characterises the EEO reaction rate. Table 2 shows the results of the analysis of the rate of gas evolution.

Table 2. The rate of gas evolution for three ethanol solutions.

Volume Ratio C_2H_5OH/H_2O	95/5	60/40	30/70
	120	540	105
t, s	360	900	870
	840	2550	2760
	19.8	4.3	22.6
$V \cdot 10^{-3}$, cm^3/s	6.6	2.6	2.6
	2.8	1	0.8

It has been found that the rate of gas evolution and the EEO is higher for solution 95/5. The high concentration of ethanol molecules promotes rapid adsorption. In addition, the rate of gas evolution is gradually reduced, which may be due to several factors:

1. solution depletion,
2. porous layer destruction,
3. contamination of the porous layer surface with reaction products.

The first two factors do not have an effect on the rate reduction. The porous layer destruction (SEM) and solution depletion has not been established. The addition of alcohol to the solution after the gas evolution stopped did not resume the process, while the as-prepared sample oxidized the spent solution. Treatment of the used sample in hydrofluoric acid contributed to the resumption of intense gas evolution in the spent solution.

Figure 8 shows SEM images of Pd/por-Si surface after EEO.

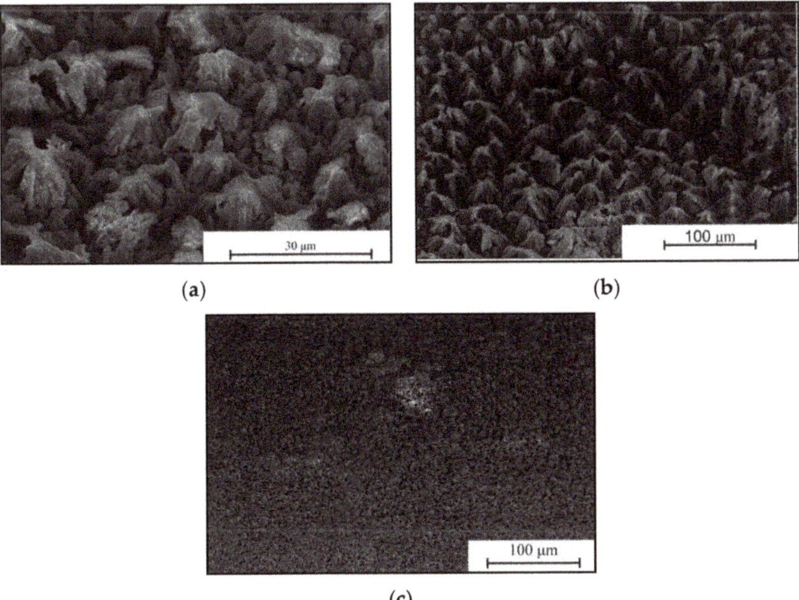

Figure 8. SEM images of porous silicon after EEO of solutions contained (a) 95/5, (b) 60/40 and (c) 30/70 ethanol/water at 25 °C during 30 min.

In Figure 8, it can be seen that the gradual reduction of the gas evolution rate was due to contamination of the surface with reaction products. The higher the concentration of ethanol in the solution, the denser the precipitates. The thickness of the precipitate layer covering the porous layer has a value of several micrometres. The element analysis (EDX method) of the porous surface after EEO (Figure 9) allows us to determine a non-uniform distribution of elements into the surface. The elements in porous silicon are Si, O, and C (Table 3). The element in porous silicon at Spots 1, 3 and 4 is silicon.

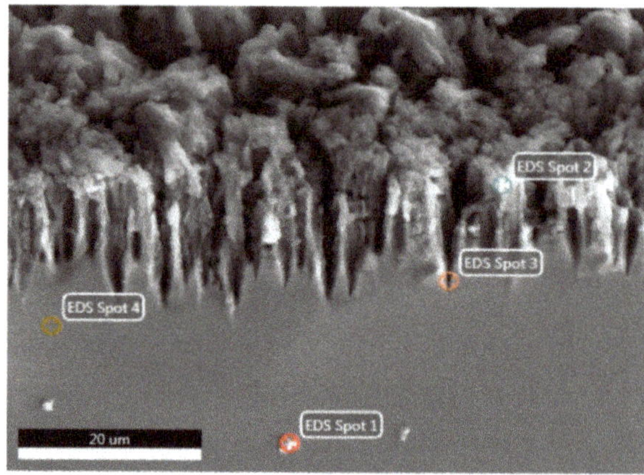

Figure 9. SEM image of porous silicon after electrooxidation and energy-dispersive X-ray (EDX) analysis spots.

Table 3. The element analysis of Spot 2 for porous silicon after 120 min electrooxidation of different solutions.

Volume Ratio C_2H_5OH/H_2O	Elements	Weight %	Atomic %
	C	2.37	5.23
95/5	O	3.33	5.53
	Si	94.30	89.23
	C	4.82	10.16
60/40	O	8.35	13.23
	Si	86.83	76.61
	C	3.20	7.05
30/70	O	2.56	4.23
	Si	94.23	88.71

Chemical bonds between the components of the porous layer were analyzed by infrared (IR) reflectance spectroscopy (Figure 10).

IR reflection spectra show the presence of bands typical for EEO by Pt and Pd catalysts. Table 4 presents wavenumbers corresponding to the bonds.

The CO_3^{2-} may be observed near 2846 cm^{-1}. The acetate was displayed as two intense peaks at 1553 and 1410 cm^{-1}. As the concentration of ethanol in solution increases, the acetate of CH_3COO^- band (1550 cm^{-1}) and ν (C–H) of CH_3CH_2OH band (2900 cm^{-1}) intensities also increase. As the concentration of ethanol in solution increases, the Si-H wag band intensities decrease. Decrease in intensity may be due to the increasing of the thickness of the precipitate layer covering the porous layer.

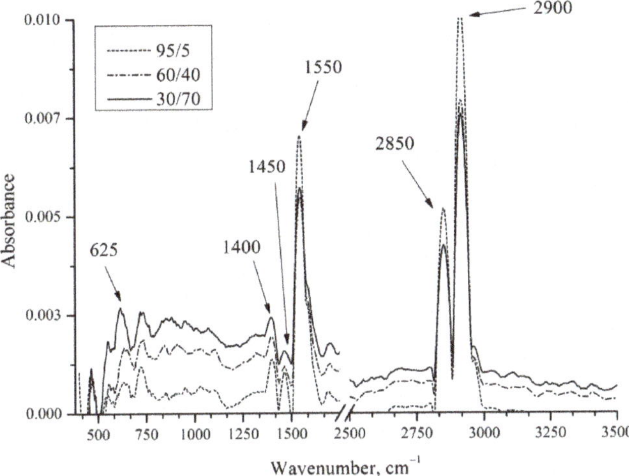

Figure 10. FTIR (Fourier-transform infrared) spectra of porous silicon after 120 min of electrooxidation of ethanol with different concentration.

Table 4. Surface bonding of porous silicon after electrooxidation (EEO).

Wave Number, cm^{-1}	Bonds, Vibration Mode
625	Si–H wag [19]
1391	δ(C–H) of CH$_3$CH$_2$OH
1400	acetate νs CH$_3$COO$^-$ [20]
1452	δ(C–H) of CH$_3$CH$_2$OH
1550	acetate νs CH$_3$COO$^-$ [20]
2850	CO$_3^{2-}$
2900	ν(C–H) of CH$_3$CH$_2$OH [21]

4. Discussion

The studies of EEO products on platinum catalysts using various analytical methods show that the reaction predominantly involves ethanol oxidation to CO$_2$ [22]:

Anodic reaction on Pd:

$$C_2H_5OH \xrightarrow{2e} CH_3OH \xrightarrow{2e} CH_X + CO \xrightarrow{8e} 2CO_2 \tag{3}$$

Cathodic reaction on Si:

$$12H^+ + 3O_2 + 12e^- \rightarrow 6H_2O \tag{4}$$

O$_2$ and H$_2$ is a result of water splitting by porous silicon because of water solutions of ethanol [23]. Figure 11 shows the mass activity as a function of time on the Pd/porous silicon with a different thickness of por-Si for the solution No. 1–3. The mass activity characterises the amount of ethanol that was oxidized by the sample over a period of time *t* and the behavior of the process. The measurement was performed in a two-electrode cell, Pd/Si- anode, Pt- cathode. When the Si/Pd-system is dipped into the ethanol solution, EEO is occurred at the Pd/Si surface. The electrons' transport is going through the porous layer and silicon wafer. The electrons diffuse into the semiconductor and accumulated at the wafer's rear (unload) side. The current can be registered in the galvanic cell (Figure 10). The current flows through the electrolyte between the Pt-cathode and Si/Pd-anode. The Pt-electrode is arranged in the electrolyte in immediate proximity to the sample surface.

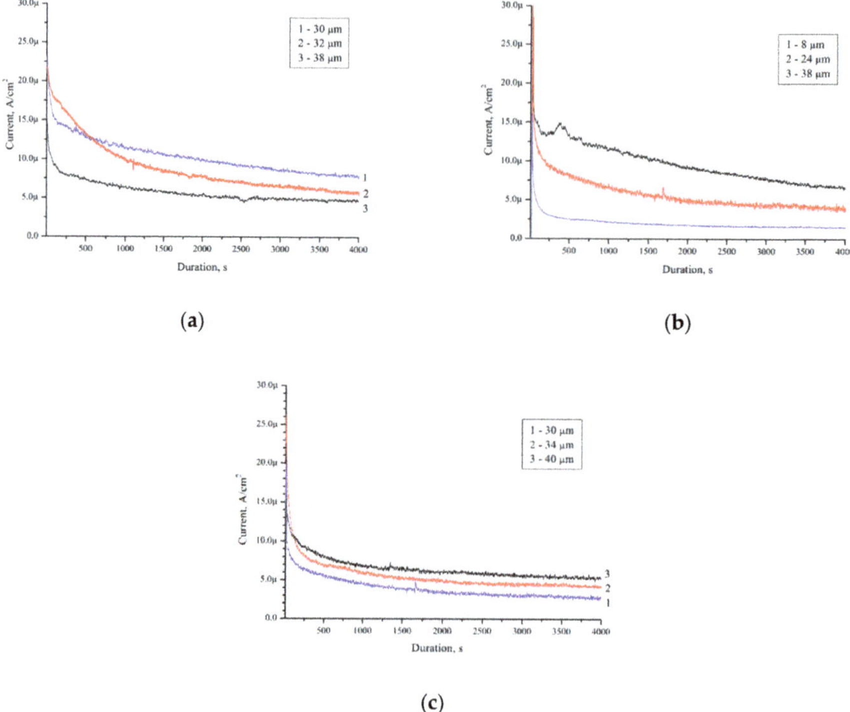

(a)

(b)

(c)

Figure 11. Current–time curves measured on Pd/por-Si for solutions: (**a**) No. 1 (10/90), (**b**) No. 2 (50/50) and (**c**) No. 3 (95/5).

The current density decreased with time for all porous samples and solutions. All catalysts showed the maximum current densities (J_{max}) immediately after the step (I section) (Figure 12). Then, the current decreased with time (II section). After a few minutes, the current achieved a pseudo-steady state (III section). The curve type corresponds to current–time curves measured during electrooxidation of dimethyl ether on Pt/C and PtMe/C catalysts in sulphuric acid [24].

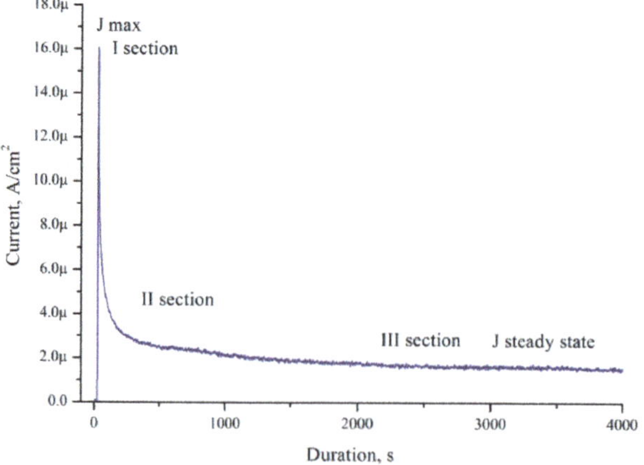

Figure 12. Typical $J(t)$ sections.

Decrease of the current density is due to the formation of the contamination on the porous silicon surface. Table 5 shows the current–time curves analysis for different solutions.

Table 5. The current–time curves analysis for different solutions.

Volume Ratio C_2H_5OH/H_2O	Porous Silicon Thickness, μm	J_{max}, μA/cm^2	$J_{steady\ state}$, μA/cm^2
10/90	30	23	8
	32	22	6
	38	17.5	5
50/50	8	16	2
	24	52.5	3.5
	38	77	7
95/5	30	54	3
	34	33	4.5
	40	79.6	5.5

It was established that J_{max} depends on the thickness of the porous silicon and concentration of ethanol. Such dependence is due to the EEO reactions yield. In this case, the thickness of the porous layer affects the amount of Pd particles in the porous layer, due to the sample preparation. The minimum value of J_{max} and $J_{steady\ state}$ is observed for the case with a minimum porous layer thickness of 8 μm. The maximum values of J_{max} equal to 77 and 79.6 μA/cm^2, and $J_{steady\ state}$ equal to 7 and 5.5 μA/cm^2, are observed for 38–40 μm thick porous layers. Ethanol concentration in solution does not affect $J_{steady\ state}$, but affects J_{max}. J_{max} reaches a value between 17 to 23 μA/cm^2 for solution 10/90. J_{max} reaches a value between 33 to 79.6 μA/cm^2 for solution 95/5. This may be due to the non-wettability of the surface of porous silicon formed Pd nanoparticles-assisted etching [25] by solution 95/5 and 50/50 (the contact angles are 140°). The contact angle for solutions 10/90 is 140°. The higher the ethanol concentration, the less the contact angle on porous surface [26]. Solution 10/90 showed the best activity for the electrooxidation in this study because of the high value of $J_{steady\ state}$. Low surface contamination during electrooxidation (Figure 8c) facilitates ethanol access to the Pd surface, intensive mass transfer, and high $J_{steady\ state}$ value.

The corresponding charges from Figure 11 can be extracted and plotted versus a time scale to indicate the rate of formation of adsorbed species at this preparation potential of Pd/por-Si. The value of $Q_{Excess\ Carrier}$ was determined by numerical integration of the dependence of the current on time and presented in Figure 13. The $Q_{Excess\ Carrier}$ value is characterising excess charge carriers diffused into the substrate during EEO.

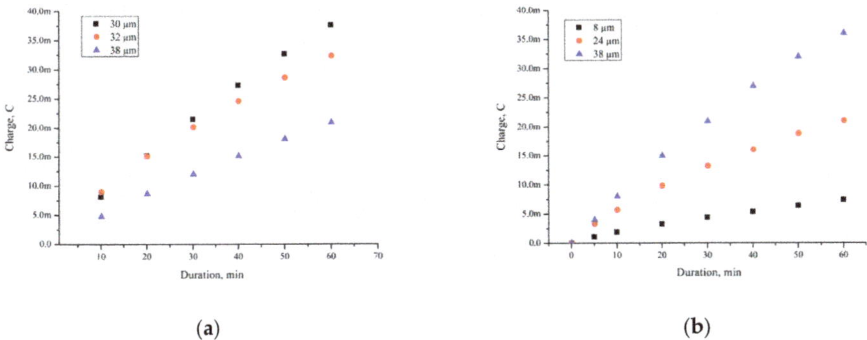

(**a**)   (**b**)

Figure 13. *Cont.*

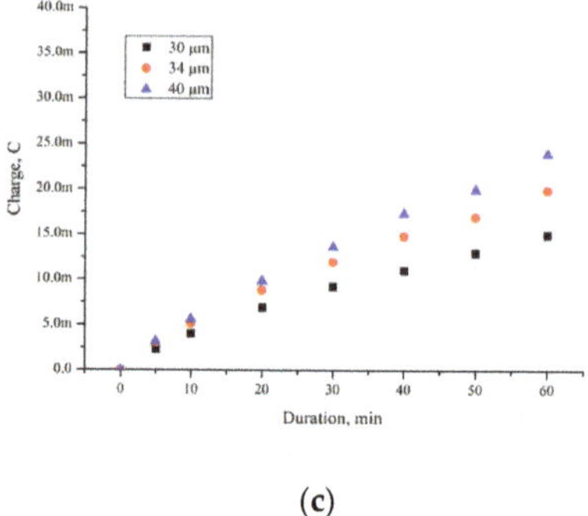

(**c**)

Figure 13. The $Q_{\text{Excess Carrier}}$ versus duration of ethanol oxidation of porous silicon with different thickness and solution: (**a**) 10/90, (**b**) 50/50, (**c**) 95/5.

The charge of time is described by a polynomial of degree 2 with $R = 0.99\%$. The charge passing through the substrate $Q_{\text{Excess Carrier}}$ depends on a few factors: concentration of charge carriers injected into silicon Q_{total}, substrate thickness and specific resistivity (Q_{sub}), and porous layer thickness ($Q_{\text{por-Si}}$).

$$Q_{total} = Q_{Excess\ Carrier} + Q_{sub} + Q_{por} \tag{5}$$

Q_{total} characterises the value of all charge carriers involved in the electrooxidation of ethanol. It the case of the present research work, Q_{total} depends on Pd/por-Si contact area. The increase in charge carrier concentration on the surface of the hole, caused by injection, leads to the appearance of a diffusion electron flow directed along the x-axis perpendicular to the semiconductor surface, with the result that the carrier concentration increases not only on the surface but also in the depth of the semiconductor. In this case, injected carriers go deeper into the semiconductor at different distances, where they are recombined.

Figure 14 shows current–time curves measured on Pd/por-Si for different solutions at 25 °C. A decrease in the wafer thickness by 70 μm, with the same thickness of porous silicon 30 μm, increases the charge from 36 to 140 mC for solution 10/90, from 20 to 100 mC for solution 50/50, and from 14 to 72 mC for solution 95/5 at 3600 s oxidation.

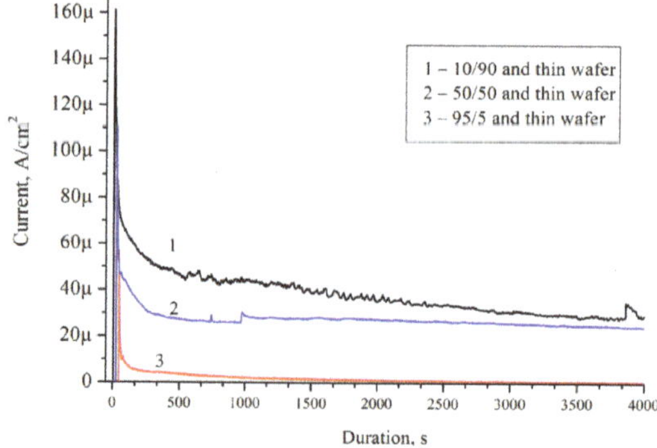

Figure 14. Current–time curves measured on Pd/por-Si for solutions: 10/90, 50/50 and 95/5.

Having excluded the contribution of the thickness and resistivity of a single crystal silicon wafer, as well as the thickness of the porous layer because of equal value, we can obtain the equation:

$$Q_{total\ n} - Q_{Excess\ Carrier\ n} = Q_{total\ m} - Q_{Excess\ Carrier\ m} \tag{6}$$

where n and m are the solutions number, and $Q_{\text{Excess Carrier}}$ is measured by short-circuit current in the galvanic cells. We can calculate Q_{total} for unknown solutions for any duration, using the same porous silicon samples and one test solution with Q_{total}.

This approach can be used for porous silicon formation by Pd nanoparticles-assisted etching. The process of forming a porous layer is identical, with the only difference being that the thickness of the initial single crystal is 525 μm (Figure 15a) and 336 μm (Figure 15b), respectively.

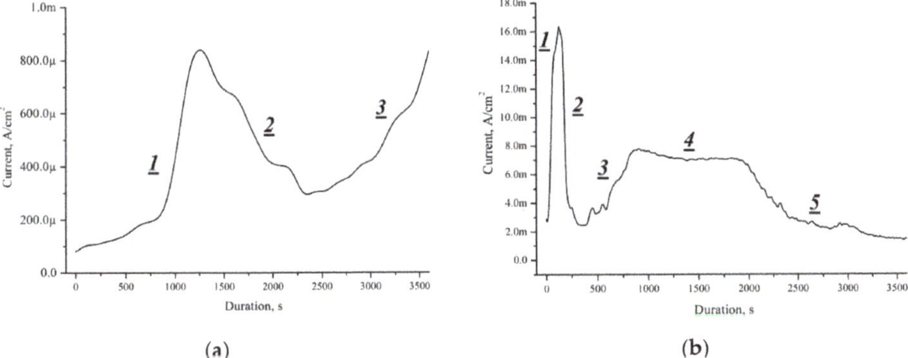

Figure 15. Current–time curves measured during Pd nanoparticles-assisted etching of silicon with thickness: (**a**) 525 μm, (**b**) 336 μm.

The $J(t)$ curves characterise the etching mechanism. Five characteristic regions can be identified:

(I)   current increasing,
(II)   slowing growth rate and subsequently decreased current,
(III) current increasing,
(IV)   constant current,
(V)   current decreasing.

$J(t)$ reflects a change of the area (S) of the electrochemical reaction front. The changes of J with time (Figure 14) are related to the evolution of the Si morphology during pore nucleation. Pore formation takes place after the immersion of silicon into the solution containing HF and H_2O_2. The pore area depends on the duration of the treatment [27]. The increase in the surface area leads to the growth of the current density, the first region.

The current growth continues until a porous layer of critical thickness is formed on the Si surface. In this case, access of the solution to the surface of monocrystalline silicon becomes limited. The transport of holes through porous silicon is difficult due to the high specific resistance of the por-Si [27]. With an increase in the thickness of the porous layer, the concentration of excess holes that are not involved in the dissolution of Si becomes smaller, the second region. A further effect of the solution on the surface leads to the dissolution of the porous layer (the beginning of region III). Dissolution of por-Si reduces the thickness of the porous layer to less than the critical value, which increases the concentration of holes in Si, and, consequently, the current in region III. The current is constant in region IV due to the growth and etching of the porous silicon. The porous layer growth and current decrease in the region V.

The charge of the first cycle (single crystal is 525 μm) takes the value of 1.5 C, while the second (single crystal is 336 μm) 17.7 C. Thus, in a single crystal 189 μm thick with a specific resistance of 0.01 Ω·cm, carriers of charge of 16.2 C recombine. Removing silicon can increase the charge passing through the sample by 11.8 times. The currents of the second cycle have values exceeding the currents of the first cycle and allow a detailed study of the structural change during Pd nanoparticles-assisted chemical etching of silicon.

5. Conclusions

It is shown that por-Si, formed by Pd nanoparticles-assisted chemical etching, has the property of ethanol electrooxidation. Intense gas evolution is observed from the metal/porous silicon immersed in ethanol solutions. The chromatographic analysis of EEO products on por-Si/Pd shows that the main products are CO_2, CH_4, H_2, CO, O_2, methanol and water vapor. The duration of the gas evolution is linearly dependent on the porosity of the layer. Therefore, the high porosity of the sample ensures access of the reactants to the surface of por-Si/Pd and removal of the reaction products. The gradual reduction of the gas evolution rate was due to contamination of the surface with reaction products. The mass activity, as a function of time, was measured by the short-circuit current in the galvanic cells.

It was established that J_{max} depends on the thickness of the porous silicon and concentration of ethanol. Such dependence is due to the EEO reactions yield. In this case, the thickness of the porous layer affects the amount of Pd particles in the porous layer, due to the sample preparation. The minimum value of J_{max} and $J_{steady\ state}$ is observed for the case with a minimum porous layer thickness of 8 μm. The maximum value of J_{max} equals to 77 and 79.6 μA/cm^2, and for $J_{steady\ state}$ equals to 7 and 5.5 μA/cm^2, is observed for 38–40 μm thick porous layers. Ethanol concentration in solution does not affect $J_{steady\ state}$, but affects J_{max}. J_{max} reaches a value between 17 to 23 μA/cm^2 for solution 10/90. J_{max} reaches a value between 33 to 79.6 μA/cm^2 for solution 95/5.

A decrease in the wafer thickness by 70 μm, with the same thickness of porous silicon, increases the charge carriers, diffused into the substrate, from 36 to 140 mC for solution 10/90, from 20 to 100 mC for solution 50/50, and from 14 to 72 mC for solution 95/5 at 3600 s oxidation. Thus, the porous silicon thickness, porosity and solution composition are the main factors defined EEO efficiency.

Author Contributions: Conceptualization, O.V.; methodology, O.V. and S.G.; validation, O.V.; formal analysis, O.V.; investigation, G.S., T.M. and A.D.; resources, O.V., T.M. and A.D.; writing—original draft preparation, O.V.; writing—review and editing, S.G. and T.M.; visualization, O.V.; supervision, S.G.; funding acquisition, O.V.

Funding: This investigation was supported by the Russian Science Foundation (project No. 19-79-00205).

Conflicts of Interest: The authors declare no conflicts of interest.

References

1. Minh, N.; Shirley Meng, Y. Future energy, fuel cells, and solid-oxide fuel-cell technology. *MRS Bull.* **2019**, *44*, 682–683. [CrossRef]
2. Murdoch, M.; Waterhouse, G.I.N.; Nadeem, M.A.; Metson, J.B.; Keane, M.A.; Howe, R.F.; Llorca, J.; Idriss, H. The effect of gold loading and particle size on photocatalytic hydrogen production from ethanol over Au/TiO$_2$ nanoparticles. *Nat. Chem.* **2011**, *3*, 489–492. [CrossRef]
3. Demirbas, A. Comparison of transesterification methods for production of biodiesel from vegetable oils and fats. *Energy Convers. Manag.* **2008**, *49*, 125–130. [CrossRef]
4. Pichonat, T.; Bernard, G.-M. Realization of porous silicon based miniature fuel cells. *J. Power Sources* **2006**, *154*, 198–201. [CrossRef]
5. Karbassian, F.; Rajabali, S.; Chimeh, A.; Mohajerzadeh, S.; Asl-Soleimani, E. Luminescent porous silicon prepared by reactive ion etching. *J. Phys. D Appl. Phys.* **2014**, *47*, 385103–385111. [CrossRef]
6. Barillaro, G.; Nannini, A.; Piotto, M. Electrochemical etching in HF solution for silicon micromachining. *Sens. Actuators A Phys.* **2002**, *102*, 195–201. [CrossRef]
7. Lehmann, V.; Föll, H. Formation Mechanism and Properties of Electrochemically Etched Trenches in n-Type Silicon. *J. Electrochem. Soc.* **1990**, *137*, 653–659. [CrossRef]
8. Gavrilov, S.A.; Karavanskii, V.A.; Sorokin, I.N. Effect of the electrolyte composition on properties of porous silicon layers. *Russ. J. Electrochem.* **1999**, *35*, 729–734.
9. Akan, R.; Parfeniukas, K.; Vogt, C.; Toprak, M.S.; Vogt, U. Reaction control of metal-assisted chemical etching for silicon-based zone plate nanostructures. *RSC Adv. J.* **2018**, *9*, 12628–12634. [CrossRef]
10. Pyatilova, O.; Gavrilov, S.; Savitskiy, A.; Dudin, A.; Pavlov, A.; Shilyaeva, Y. Investigation of porous Si formed by metal-assisted chemical etching with Au as catalyst. *J. Phys. Conf. Ser.* **2017**, *829*, 1. [CrossRef]

11. Kumar, J.; Ingole, S. Effect of Silicon Conductivity and HF/H$_2$O$_2$ Ratio on Morphology of Silicon Nanostructures Obtained via Metal-Assisted Chemical Etching. *J. Electron. Mat.* **2018**, *47*, 1583–1588. [CrossRef]

12. Yae, S.; Morii, Y.; Fukumuro, N.; Matsuda, H. Catalytic activity of noble metals for metal-assisted chemical etching of silicon. *Nanoscale Res.* **2012**, *7*, 352–357. [CrossRef]

13. Zhu, B.; Lund, P.; Raza, R.; Ma, Y.; Fan, L.; Afzal, M.; Patakangas, J.; He, Y.; Zhao, Y.; Tan, W.; et al. Schottky Junction Effect on High Performance Fuel Cells Based on Nanocomposite Materials. *Adv. Energy Mat.* **2015**, *5*, 1401895–1401901. [CrossRef]

14. Dzhafarov, T.; Yuksel, S.; Aydin, M. Nanoporous Silicon-Based Ammonia-Fed Fuel Cells. *Mater. Sci. Appl.* **2014**, *5*, 1020–1026. [CrossRef]

15. Liang, Z.; Zhao, T.; Xu, J.; Zhu, L. Mechanism study of the ethanol oxidation reaction on palladium in alkaline media. *J. Electrochem. Acta* **2009**, *54*, 2203–2208. [CrossRef]

16. Farsadrooh, M.; Torrero, J.; Pascual, L.; Peña, M.A.; Retuerto, M.; Rojas, S. Two-dimensional Pd-nanosheets as efficientelectrocatalysts for ethanol electrooxidation. Evidences of the C-C scission at low potentials. *Appl. Catal. B Environ.* **2018**, *237*, 866–875. [CrossRef]

17. Ju, K.S.; Pak, S.N.; Ri, C.N.; Ryo, H.S.; Kim, K.-I.; So, S.-R.; Ri, C.-K.; Ri, S.-P.; Nam, K.-W.; Pak, K.-S.; et al. Ethanol electro-oxidation on carbon-supported Pt1Mn3 catalyst investigated by pinhole on-line electrochemical mass spectrometry. *Chem. Phys. Lett.* **2019**, *727*, 78–84. [CrossRef]

18. Backes, A.; Leitgeb, M.; Bittner, A.; Schmid, U. Temperature Dependent Pore Formation in Metal Assisted Chemical Etching of Silicon. *J. Solid State Sci. Technol.* **2016**, *5*, 653–656. [CrossRef]

19. Canham, L. *Handbook of Porous Silicon*; Springer: Basel, Switzerland, 2018.

20. Christensen, P.A.; Jones, S.W.M.; Hamnett, A. In Situ FTIR Studies of Ethanol Oxidation at Polycrystalline Pt in Alkaline Solution. *J. Phys. Chem. C* **2012**, *116*, 24681–24689. [CrossRef]

21. Iwasita, T.; Pastor, E. A dems and FTir spectroscopic investigation of adsorbed ethanol on polycrystalline platinum. *Electrochim. Acta* **1994**, *39*, 531–537. [CrossRef]

22. Teng, X. Anodic Catalyst Design for the Ethanol Oxidation Fuel Cell Reactions. *Mater. Sci.* **2013**, 473–484.

23. Dai, F.; Zai, J.; Yi, R.; Gordin, M.L.; Sohn, H.; Chen, S.; Wang, D. Bottom-up synthesis of high surface area mesoporous crystalline silicon and evaluation of its hydrogen evolution performance. *Nat. Commun.* **2014**, *5*, 3605. [CrossRef]

24. Liu, Y.; Mitsushima, S.; Ota, K.-I.; Kamiya, N. Electro-oxidation of dimethyl ether on Pt/C and PtMe/C catalysts in sulfuric acid. *Electrochim. Acta* **2006**, *51*, 6503–6509. [CrossRef]

25. Volovlikova, O.V.; Gavrilov, S.A.; Silakov, G.O.; Zheleznyakova, A.V.; Dudin, A.A. Preparation of Hydrophobic Porous Silicon by Metal-assisted Etching with Pd-Catalyst. *Russ. J. Electrochem.* **2019**, *55*, 1186–1195.

26. Volovlikova, O.V.; Silakov, G.O.; Gavrilov, S.A.; Dudin, A.A.; Diudbin, G.O.; Shilyaeva, Y.I. Investigation of the Morphological Evolution and Etching Kinetics of black Silicon During Ni-Assisted Chemical Etching. *J. Phys. Conf. Ser.* **2018**, *987*, 012039. [CrossRef]

27. Bisi, O.; Ossicini, S.; Pavesi, L. Porous silicon: A quantum sponge structure for silicon based optoelectronics. *Surf. Sci. Rep.* **2000**, *38*, 1–126. [CrossRef]

 © 2019 by the authors. Licensee MDPI, Basel, Switzerland. This article is an open access article distributed under the terms and conditions of the Creative Commons Attribution (CC BY) license (http://creativecommons.org/licenses/by/4.0/).

Review

Microfabrication of X-ray Optics by Metal Assisted Chemical Etching: A Review

Lucia Romano [1,2,3,*] and Marco Stampanoni [1,2]

1   Institute for Biomedical Engineering, ETH Zürich, 8092 Zürich, Switzerland; marco.stampanoni@psi.ch
2   Paul Scherrer Institut, Forschungsstrasse 111, CH-5232 Villigen, Switzerland
3   CNR-IMM, Department of Physics, University of Catania, 64 via S. Sofia, 95123 Catania, Italy
*   Correspondence: lucia.romano@psi.ch or lucia.romano@ct.infn.it

Received: 14 May 2020; Accepted: 10 June 2020; Published: 12 June 2020

Abstract: High-aspect-ratio silicon micro- and nanostructures are technologically relevant in several applications, such as microelectronics, microelectromechanical systems, sensors, thermoelectric materials, battery anodes, solar cells, photonic devices, and X-ray optics. Microfabrication is usually achieved by dry-etch with reactive ions and KOH based wet-etch, metal assisted chemical etching (MacEtch) is emerging as a new etching technique that allows huge aspect ratio for feature size in the nanoscale. To date, a specialized review of MacEtch that considers both the fundamentals and X-ray optics applications is missing in the literature. This review aims to provide a comprehensive summary including: (i) fundamental mechanism; (ii) basics and roles to perform uniform etching in direction perpendicular to the <100> Si substrate; (iii) several examples of X-ray optics fabricated by MacEtch such as line gratings, circular gratings array, Fresnel zone plates, and other X-ray lenses; (iv) materials and methods for a full fabrication of absorbing gratings and the application in X-ray grating based interferometry; and (v) future perspectives of X-ray optics fabrication. The review provides researchers and engineers with an extensive and updated understanding of the principles and applications of MacEtch as a new technology for X-ray optics fabrication.

Keywords: X-ray grating interferometry; catalyst; silicon; gold electroplating

1. Introduction

High resolution and high-efficiency diffractive optics have largely been unavailable for hard X-rays where many scientific, technological, and biomedical applications exist. This is due to the long-standing challenge of fabricating high aspect ratio high-resolution micro- and nano-structures.

Fabrication of high-aspect-ratio silicon micro- and nano-structures is a key process in many applications, such as microelectronics [1], microelectromechanical systems [2,3], sensors [4], thermoelectric materials [5], battery anodes [6], solar cells [7], photonic devices [8], and X-ray optics [9]. Microfabrication is usually achieved by reactive ion etching [10], which requires high investment in tools and maintenance. KOH-based wet etching [11,12] has been used for microfabrication in Si at micro- and nano-scale. However, the aspect ratio of etched trenches is limited by the etching rate ratio between different crystallographic orientations and only possible in simple geometries like linear gratings or crossed linear gratings defined by the direction of <111> crystallographic planes of Si. As an alternative approach for fabricating Si microstructures, metal assisted chemical etching [13] has attracted great interest [14] because of its simplicity, low fabrication costs, and ability to generate high aspect ratio nanostructures such as nanowires [15]. Several acronyms were reported for this process—MACE, MAE, MacEtch—since 2015 the community seemed to agree with the common acronym of "MacEtch", which was firstly introduced by X. Li [16] to distinguish the unique properties with respect of standard wet-etch and dry-etch techniques. Unlike KOH wet-etch [12], the MacEtch process is almost independent of crystal orientation and may be used to create a wide variety of patterns, without suffering of microloading effects of dry-etch. An advantage of the method is the considerable reduction in fabrication costs and complexity with respect to the other techniques. MacEtch fabrication of nanoscale patterns has been successfully applied for synchrotron-based X-ray imaging methods [17,18]. For X-ray grating interferometry imaging, the fabrication of Si micrograting requires sharp vertical profiles, high aspect ratios, high accuracy of pitch size and duty cycle, uniformity over large area, and, finally, the possibility to fill up the Si template with a high X-ray absorbing material [19,20] such as gold [21,22]. These requirements are especially stringent for X-ray medical diagnostics for which extremely large field of view is necessary. Thus gratings require microfabrication on area of many squared centimeters [9], with aspect ratio and pitch size that depend on the used energy, specific design and performances (pitch size in the range of 1–20 μm, aspect ratio in the range of 10–100).

Since its discovery in 2000, by Li et al. [13], MacEtch of silicon has emerged as a new technique capable of fabricating 3D nano- and micro-structures of several shapes and applications [23]:—nano-porous film, nanowires [24], 3D objects [25], trenches, vias [26], micro-fins [27], nano-scale grooves, surface antireflection texturing [28], optoelectronic devices such as solar cells [29] and photodetectors [30], sensor devices [31], X-ray optics—in a few semiconductors substrates: Si [15], Ge [30], poly-Si [32], GaAs [33], β-Ga$_2$O$_3$ [27], SiC [34], etc.—and different catalysts: Ag, Au, Cu, Pt, and Pd [15]. MacEtch has been developed with a strong controlled vertical directionality with respect to the substrate and successfully applied for producing X-ray zone plates [17,23,35–38] and diffraction gratings [19,20,37,39–41]. In MacEtch, a catalyst layer (e.g., Au) is patterned onto the substrate (e.g., Si) to locally increase the dissolution rate of the substrate material in an etchant solution including a fluoride etchant such as hydrofluoric acid (HF) and an oxidizing agent such as hydrogen peroxide (H$_2$O$_2$).

To date, no comprehensive review of patterned microstructures by MacEtch exists in the literature. The existing reviews on MacEtch rarely focus on the aspects of X-ray gratings fabrication. This paper provides an extensive overview of the fundamentals and recent developments of MacEtch as well as addressing the research gaps in this field. After an overview about the MacEtch mechanism, we dedicated a particular attention to the conditions (catalyst, additives, and reaction temperature) to ensure the vertical etching of the (100) Si substrates. Then, we described the procedures for gratings fabrication, from pattern design to transfer in the silicon substrate and finally the template filling with a high X-ray absorbing material such as Au. In the last section we discussed the recent applications of Fresnel zone plates and X-ray interferometric gratings fabricated by MacEtch. In the concluding remarks we outlined the major challenges for large-scale MacEtch X-ray optics and the perspectives of MacEtch microfabrication.

MacEtch offers the possibility to fabricate high aspect ratio structures for hard X-ray diffractive optics and opens up new opportunities for high resolution imaging with compact X-ray sources and for synchrotrons and X-ray-free electron lasers with more complex wave front manipulation.

2. MacEtch Mechanism

The mechanism of MacEtch has been extensively debated in literature [15] even with controversial interpretations [42]. Etching occurs when the metal (catalyst) patterned Si substrate is immersed in a solution with an etchant (for example HF) and an oxidizer (for example H$_2$O$_2$). The solution-metal-silicon system constitutes a microscopic electrochemical cell that induces anodic silicon etch.

For the most commonly employed oxidants (H$_2$O$_2$), the proposed cathodic reactions provide free positive carriers to be transferred to the silicon, according to Equation (1). In the anode reaction, the silicon consumes the positive carriers and is solubilized through oxidation (Equation (2)). The concentration of holes becomes higher in the region surrounding the metal catalyst, where silicon is readily oxidized by HF and forms silicon fluoride.

$$H_2O_2 + 2H^+ \rightarrow 2H_2O + 2h^+ \tag{1}$$

$$Si + 4h^+ + 6HF \rightarrow SiF_6^{2-} + 6H^+ \tag{2}$$

Common oxidizers for MacEtch and their associated cathode reactions are reviewed by Chiappini et al. [43]. A wide variety of metal salts can induce silicon porosification without the addition of any other oxidizer, as their electrochemical potential is sufficiently high to directly inject holes in the valence band of silicon [43]. Several other oxidizing agents have been studied [44], including oxygen [37,45] with the following reaction (Equation (3)):

$$O_2 + 4H^+ + 4e^- \rightarrow 2H_2O. \tag{3}$$

Figure 1 shows Scanning Electron Microscopy (SEM) images of MacEtch after few seconds of etching. The removal of Si atoms occurs faster at the interface with the metal catalyst, where positive carriers have the maximum concentration. As the reaction proceeds, the catalyst sinks into the substrate and progressively the catalyst nanopattern is transferred to the substrate. The process continues as long as the etchants are present in the solution and the reaction byproducts diffuse out of the pattern. Despite of the simple mechanism, the full process has indeed a complex dynamic where several phenomena—mass transport of etchants and byproducts, charge carrier diffusion, catalyst stability, and gas release—interplay to determine the etching rate, the etching direction, and the quality of the etched structure.

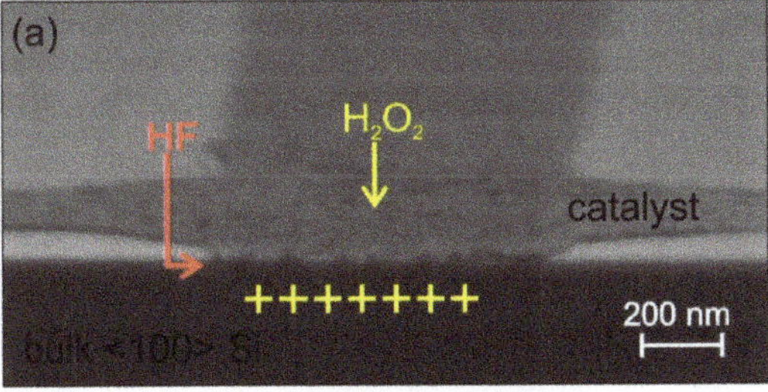

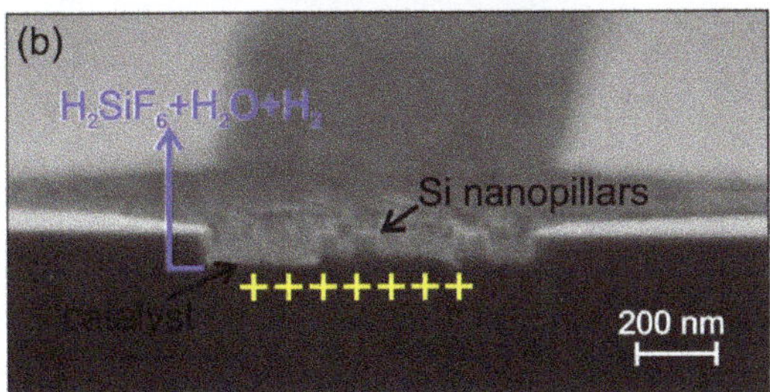

Figure 1. MacEtch mechanism in solution of HF and H_2O_2. (**a**) The metal catalyst deposited on a Si <100> substrate decomposes H_2O_2 with consequent injection of holes (+) into the semiconductor. (**b**) Si consumes the positive carriers Si, it is readily oxidized by HF and forms silicon fluoride, the process continues and the catalyst progressively sinks into Si along the <100> direction, transferring the nanostructure pattern to the Si (formation of nanopillars in this case). Images are cross-section Scanning Electron Microscopy (SEM) of nano-patterned Pt on Si (**a**) after few seconds of MacEtch in solution of HF and H_2O_2.

The etching mechanism and the composition dependence have been extensively reported in literature [15,43,46].

Hydrogen peroxide is by far the most commonly employed oxidizer in MacEtch. Chartier et al. assume that the relative concentration of HF and H_2O_2 in a MacEtch etch solution plays a similar role to the current density J_{ps} in anodic etch [46]. MacEtch solution is usually described in terms of concentrations ratio between HF and H_2O_2, according to Chartier's formula (Equation (4))

$$\rho = \frac{[HF]}{[HF] + [H_2O_2]} \tag{4}$$

where [HF] and [H_2O_2] are the molar concentration of HF and H_2O_2, respectively and Hildreth's [47] compact expression ρ[HF].

3. Vertical Etching

Huang et al. [48] demonstrated that MacEtch is intrinsically anisotropic along the preferred crystallographic <100> directions. Such an orientation dependence is related to the silicon lattice configuration at the reaction site. Removal of oxidized silicon by HF is associated with the cleavage of its back bonds, of which effective number density in different crystal planes increases with the order (100) < (110) < (111) [49]. Due to the different back-bond strength, the Si atom on the (100) surface plane is the most easily removed, and the etching occurs preferentially along the <100> directions. The anisotropy could be reduced or eliminated by varying the concentration of the etchants. In MacEtch of Si, the movement of the etching front (i.e., metal/Si interface) is a net consequence of the following two competing events: (1) injection of a positive charge carriers into bulk Si through the metal-Si interface and (2) removal of oxidized Si by HF from just underneath the catalyst metal. Since the generation of holes is related to the catalytic decomposition of H_2O_2 at the interface between the solution and the catalyst metal surface, the amount of holes injected into Si is proportional to the H_2O_2 concentration in the solution and the catalyst activity. In conditions of low H_2O_2 concentration, hole injection into Si atoms will be localized at the (100) plane, where there are the fewest Si back bonds to break, resulting in etching along the <100> direction. As the concentration of H_2O_2 increases sufficiently, removal of oxidized Si would be kinetically favored in the crystal planes with a higher density of silicon back bonds, resulting in etchings along non-<100> directions. The same argument can be played considering HF, in conditions of low HF concentration, the removal of oxidized silicon would control the reaction so the favorite etching direction is again the <100>. While, for high HF concentration, also the other directions are favored. A schematic is reported in Figure 2. It must be noted that H_2O_2 and HF are correlated in the solution, so the transition between one etching direction to the other should be determined as a function of the specific solution and the used catalyst. A complex ternary graph would result taking into account the water dilution, as reported by J. Kim et al. [50].

During the process of etching, a triggering event that produces unequal etch rates might occur. These events can change the effective forces on the catalyst and produce a resultant torque on the catalyst. [51]. Such triggering events are made more frequent in the case of higher etch rates as brought about by higher oxidant concentration or by etching at elevated temperature. To date, this represents one of the major challenge to optimize MacEtch as a reliable and controllable process for large area patterning of high aspect ratio structures. Hildreth et al. [25,52] demonstrated that controlled 3D motion of catalyst patterns during MacEtch can be achieved by locally pinning them with an electrically insulating material prior to etching. However, due to this movement, the aspect ratio achievable for features perpendicular to the substrate in an arbitrary dense pattern is limited.

Moreover, charge carriers are injected into Si and charge distribution affects the catalyst movement [53], so that parallel and elongated structures [17] are more difficult to etch than spaced cavities [26]. Several approaches have been attempted in literature to force the uniform etching along the <100> and to minimize the etching along the other directions [51] in order to realize high aspect ratio structures perpendicular to the (100) substrate (vertical etching). Electron-hole concentration balancing structures were used to achieve a vertical etch profile in X-ray zone plates [17]. Figure 3 shows some examples of balancing structures [36] used to define the vertical etching at the borders of the X-ray lens structures [41].

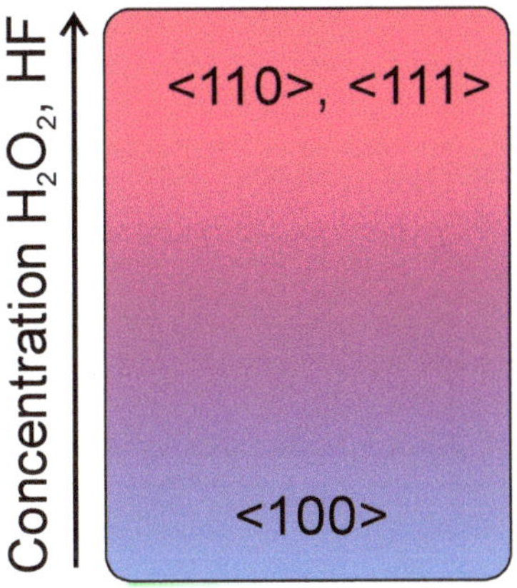

Figure 2. Schematic of preferred etching direction as a function of H_2O_2 and HF concentration. The Si back bonds are preferentially removed along the <100> directions in conditions of both low H_2O_2 and HF concentrations.

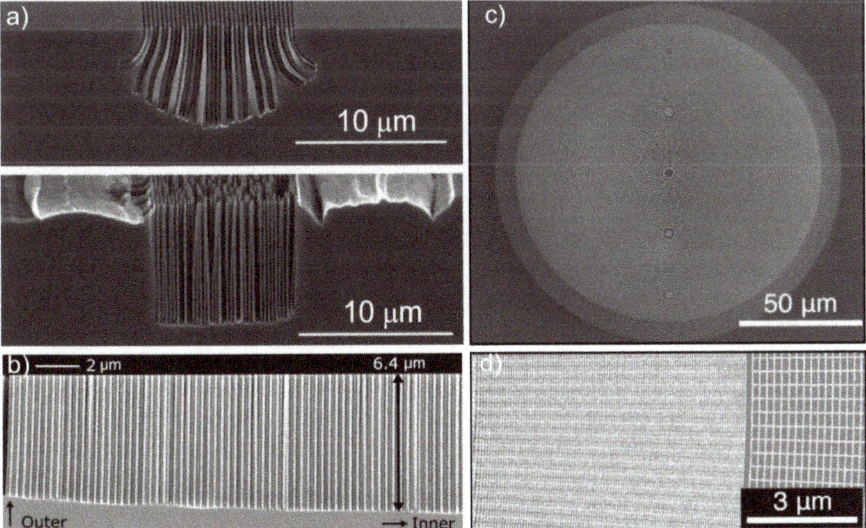

Figure 3. Balancing structures to control the vertical etching at the device border: (**a**) cross section SEM of etched linear grating (pitch 250 nm) with and without balancing structures, the figure was adapted with permission from C. Chang et al., 2014. [17]; (**b**) cross-section SEM of etched kinoform lens with outmost zone of pitch 150 nm, the figure was adapted with permission from M. Lebugle et al., 2018. [41]; (**c**) SEM of Au pattern of zone plates with balancing ring, the figure was adapted with permission from K. Li et al., 2017. [36]; and (**d**) detail of border with balancing ring in (**c**), the figure was adapted with permission from K. Li et al., 2017. [36].

Negative carbon mask [54,55], electrical bias [32,55], and magnetic catalyst [56] have been proposed to force the vertical etching and to improve the control of the catalyst movement [47,52,57]. MacEtch resulted to be very efficient for Si nanostructures, nanowires, and ordered nanopillars [15,29,58–61], but etching in the microscale regime is more critical [62], the etching rate is limited by the reactant diffusion through the metal mask. The effective transfer of reactants and their by-products would not be identical where the metal pattern size is nanometers or few micrometers. Therefore, two regimes can be distinguished in literature [2,14,63]: (i) nanoscale patterns, in which the etchant species diffuses through the pattern edges and (ii) microscale patterns with nano-porous films, in which the porosity of the film itself controls the diffusion length. In both regimes, the catalyst geometry significantly affects the etching performance. Catalyst optimization and etching conditions are here reviewed in order to address the vertical etching.

3.1. Catalyst

A wide range of transition metals can catalyze MacEtch. Noble metals are especially favored for the formation of nanowires as well as for nanostructures with defined cross sections since they better preserve their structure during the etch, as they do not dissolve in HF. Non-noble transition metals have been mostly used to form nano-pores, porous Si, and polished surfaces [43]. The most used MacEthc catalyst is Ag [15]. However, X-ray optics fabrication requires sophisticated patterning techniques such as electron beam lithography or UV photolithography and the catalyst film is usually deposited by thin film evaporation. Silver oxidation is quite difficult to prevent during thin film physical deposition, so Au is the most studied catalyst for thin film deposition. Here, we review the catalyst that have been used for X-ray optics fabrication, which are Au [17,39,40] and Pt [37]. Platinum is the metal with the highest catalytic activity so it allows to obtain the highest MacEtch rate [64]. The patterning of nanostructures requires high precision pattern transfer and high lateral resolution during etching, with MacEtch in liquid this corresponds to a condition of very high HF concentration [17]. Gold catalyst suffers of bad adhesion on silicon substrates, yet a detrimental pattern peel-off has been reported during MacEtch in conditions of high HF concentration [50,65]. On the other hand, uniform high aspect ratio has been reported for nanoporous Au catalyst in conditions of low HF and high H_2O_2 concentration [39,66]. In these conditions, the etching is more isotropic [63], the top of the trenches appear wider with respect to the bottom compromising the fidelity of the pattern transfer in the lateral dimension, so the process is not suitable for high aspect ratio structures.

Porous catalyst film is reported [2,63,66,67] to improve the etching performances of micro-scaled Si trenches structures with interconnected catalyst pattern. The porous morphology of the film allows the MacEtch reactants to pass through the catalyst spacing, significantly improving the mass transport and uniformity, which ensures a highly uniform etch rate over all the catalyst area. We recently applied the thermal de-wetting technique to carefully design the film porosity of Au and Pt catalyst and control the vertical etching in micrometer patterns of MacEtch for grating fabrication [21,37,39,40]. Thermal de-wetting is much more robust than evaporation rate to control the film morphology. De-wetting occurs when a thin metal film on a solid substrate is heated, inducing breaking and reassembling of the film [68,69]. The film morphology can be tuned as a function of film thickness and annealing temperature. Figure 4 reports an example of de-wetting for Au film and Pt film deposited on a Si substrate with a cleaned native oxide (oxygen terminated surface). The Pt de-wetting occurs in agreement with literature [69] with a progressive increase of film fractures density (250–350 °C) and finally the hole formation appeared (400–500 °C), followed by a coalescence process of holes expansion (550–600 °C). The thermal treatment in the case of Pt film has two different functions: it creates the porous structure in the metal coating and it forms a platinum silicide at the interface with the substrate that helps to stabilize the catalyst during etching [37].

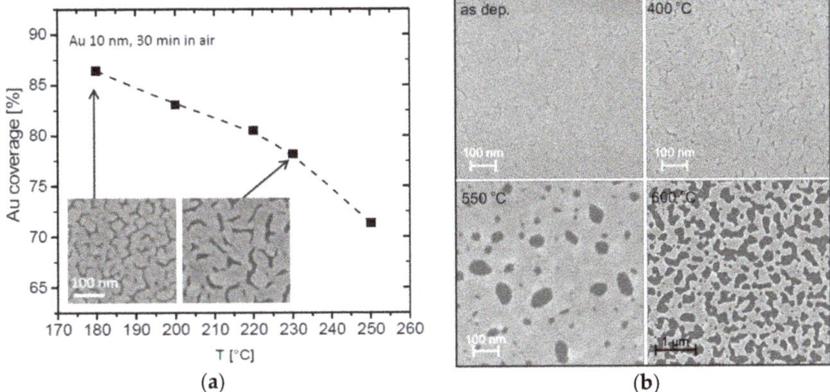

Figure 4. (**a**) Au coverage in percentage of the surface area measured in SEM images in plan-view as a function of the annealing temperature. The Au film thickness was 10 nm and the annealing was performed in air for 30 min. Insets show SEM images of Au film annealed at 180 °C (left) and 230 °C (right), the scale marker is the same in both images. The figure was adapted with permission from L. Romano et al., 2017. [40] (**b**) Pt de-wetting (12 nm) on (100) Si substrate at temperature of 400, 550, and 600 °C. The figure was adapted with permission from L. Romano et al., 2020. [21].

3.2. Alcohols Additives

Ethanol [8] and isopropanol [39] alcohols have been largely used as surfactant in MacEtch solutions. Like in KOH aqueous solutions with addition of alcohol [70], also for MacEtch the alcohol does not take directly part in the etching process, but it strongly affects the etching. Both etch rate and roughness of the etched surface depend on the alcohol concentration in the etching solution, which is connected with the adsorption phenomena on the etched surface [40]. A common issue of MacEtch is the H_2 gas release during the etching process. The H_2 is produced as a by-product of reaction [15] and it can substantially affect the etching results since very large bubbles can be formed on the surface of the grating, dramatically preventing a uniform etching. This phenomenon appeared to be much more critical in patterned microstructures than mesh pattern for nanowires since the gas bubbles can be stabilized in the etched structure with liquid solution exhibiting the Cassie-Baxter wetting state [71,72]. The surfactant forms a layer physically covering the surface and prevents the formation of large H_2 bubbles, reducing the amount and the size of etchant inhomogeneity in contact with the surface [40]. An example of grating fabricated with and without surfactant additive in the etching solution is showed in Figure 5.

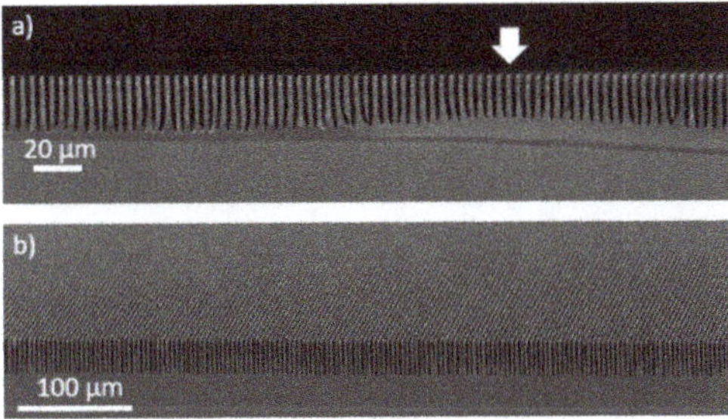

Figure 5. SEM in cross-section of 4.8 μm pitch grating etched with regular MacEtch (**a**) and MacEtch solution with the addition of isopropanol alcohol (**b**). The arrows indicate the presence of a gas bubble preventing the uniform etching of the grating. The figure was adapted with permission from L. Romano et al., 2017. [40].

Formation of porous Si is a well-known phenomenon which has been observed in MacEtch. The porous morphology of Si using MacEtch has been attributed to the diffusion of holes outside of the metal-semiconductor interface and causing an additional but reduced extent of etching in the areas outside the metal mesh pattern [73]. Depending on etching conditions, pores with different density and thickness can be found at the catalyst/Si interface, along the sidewalls, and within the etched Si nanostructures. In general, a higher oxidant concentration or higher Si doping concentration results in higher levels of porosity. Once the oxidant is reduced on the surface of noble metal, holes are injected into the Si substrate. The holes diffuse from the Si under the noble metal to the off-metal areas that may be etched and form microporous Si. Balasundaram et al. [73] showed that porosity depends on Si doping, the dopant atoms are thermodynamically favorable sites for the formation of pores, and heavily doped Si in liquid MacEtch produces very porous structures even in conditions of very low H_2O_2 concentration. The thickness of the microporous Si can be additionally reduced by adding a small amount of alcohol to the etching solution [40]. Figure 6 reports a magnified SEM of the top Si lamellas in Pt-MacEtch with additional methanol, the microporous thickness is less than 50 nm. Methanol is less affecting the etching rate with respect of isopropanol and ethanol alcohols [74].

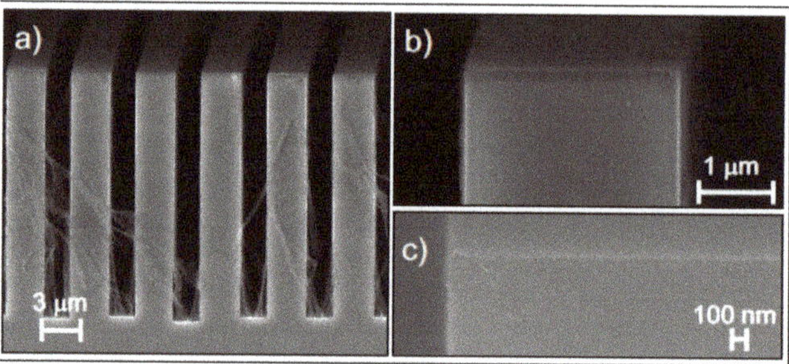

Figure 6. SEM in cross-section of Pt assisted chemical etching of silicon grating with 4.8 μm pitch (**a,c**), magnified view of the top (**b**) with etching solution $\rho(HF) = 0.99^{20}$. Magnified view of top (**c**) with additional methanol in the etching solution. The figure was adapted with permission from L. Romano et al., 2020. [21].

However, a large amount of additive (methanol, ethanol, isopropanol, and acetonitrile in the $HF–H_2O_2–H_2O$ solution) can cause the changing of the etching direction, inducing the formation of curved or tilted structures [74]. An example is reported in Figure 7, where curved Si nanowires are produced in a solution with isopropanol and acetonitrile. H_2O_2 is more severely shielded from reaction sites by the additive than HF is, due to the higher surface tension of H_2O_2, effectively increasing the HF to H_2O_2 ratio locally.

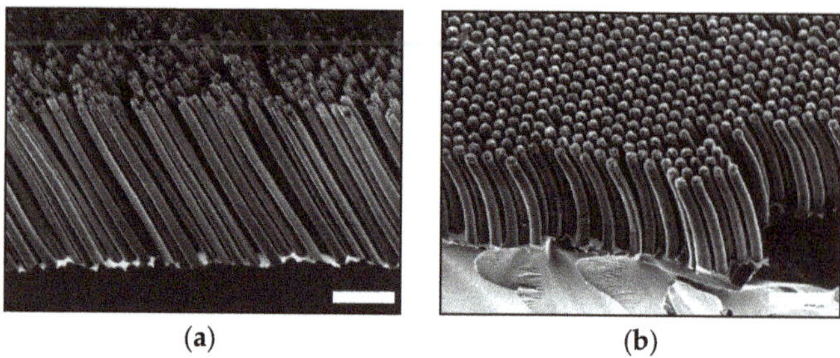

(**a**)　　　　　　　　　　　(**b**)

Figure 7. SEM images of Si nanowires etched for 20 min in $HF–H_2O_2$–water–co-solvent (**a**) isopropanol 2:1:5:2 and (**b**) acetonitrile 2:1:5:2. All scale bars are 2 μm. The figure was adapted with permission from Y. Kim et al., 2013. [74].

3.3. Temperature

With a temperature in the range of 0 °C to 50 °C, Cheng et al. [75] observed a linear relationship between length of nanowire and etching time. The etching rate increased with increasing etching temperature with an activation energy of 0.36 eV for the formation of Si nanowires on a (100) Si substrate in $AgNO_3$ and HF aqueous solution.

Temperature has also a strong effect on the etching direction. Figure 8 shows the effect of temperature on etching direction for (100) substrates, as the temperature increases (50–70 °C) the etching along the (110) instead of (100) is preferred. Figure 9 shows a comparison of etching at 30 °C and 8 °C, indicating the temperature reduction as a possible way to control the vertical etching.

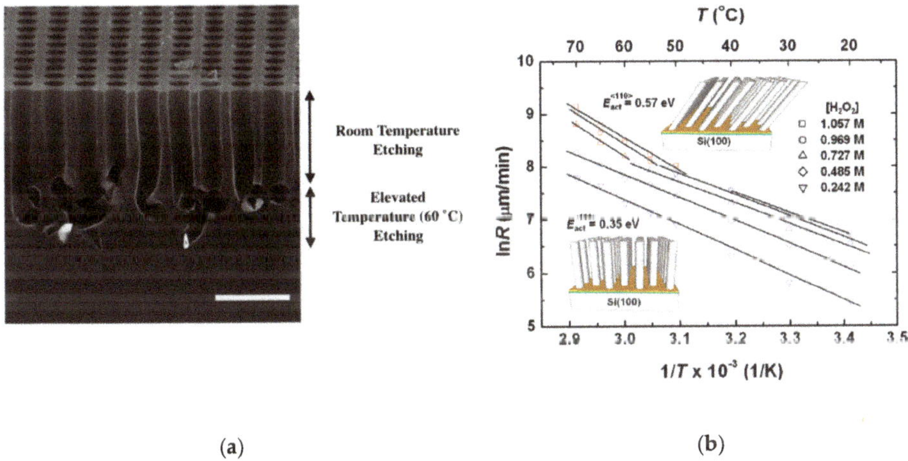

(a)   (b)

Figure 8. (**a**) Cross-sectional SEM images of the as-etched Si produced with Au catalyst of *p*-type Si (100) substrate (0.005 Ωcm) in solution of 13.5 M HF and 0.16 M H_2O_2 at room temperature for 30 min and subsequently etched at the same condition at 60 °C for 1 min. Scale bar is 2 µm. The figure was adapted with permission from L. Kong et al., 2017. [51]. (**b**) Temperature dependence of etching rate for different H_2O_2 concentration, at high temperature the etching direction changes from (100) to (110). The figure was adapted with permission from J. Kim et al., 2011. [50].

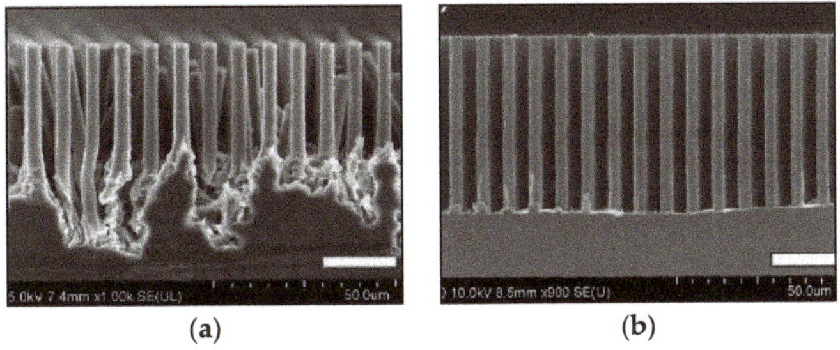

(a)   (b)

Figure 9. Cross-sectional SEM images of the as-etched Si produced with a catalyst of 40/5 nm thick Au/Ti of n-type Si (100) substrate (0.9–1.1 Ωcm). The etching temperatures are 30 °C (**a**) and 8 °C (**b**), respectively. The etching duration and concentration of H_2O_2 are 12 h and 0.2 M, respectively. Scale bar is 20 µm in all figures. The figure was adapted with permission from J. Yan et al., 2016. [76].

The Si porosity is strongly affected by the etching temperature. Excess holes tend to diffuse laterally, resulting in lateral etching and the formation of pits on the sidewalls. Cold etching temperature is also highly advantageous to reduce the pits on the sidewalls [77]. K. Balasundaram et al. [73] noted that when the etching temperature decreases, the porosity is reduced. R. Akan et al. [35] reported an increase in surface roughness and porosity at the etching temperature of 40 °C.

Another way to observe the effect of temperature is MacEtch in gas phase [37,78], where the HF is delivered in vapor phase and the oxidant in gas phase (oxygen from air) to the metal patterned Si substrate. It has been recently demonstrated that etching in the vapor phase avoids the issues related to wet-etching such as the nanostructures stiction due to capillary effects during liquid drying. Moreover, the pattern transfer from the metal mask to the silicon template is much more precise and defect-less due to the microporosity reduction and the extremely high concentration of HF, which are not accessible in wet-etching. By increasing the temperature in the range 35–40 °C, the etching rate increases in agreement with previous studies on MacEtch kinetics in liquid [75]. The etching rate has a maximum at 40 °C (see [37]), then it decreases as a function of temperature, indicating that the reaction rate is limited by the desorption of HF. Some examples of nano- and micro-structures of X-ray optics [37] are reported in Figure 10. The etching was realized by evaporating water diluted HF (50 wt.%) at room temperature and exposing the Pt-patterned Si substrate to the HF vapor and air, the gaseous O_2 present in the air worked as oxidant for the MacEtch reaction. The Pt-patterned Si substrate is held at 55 °C during the etching in order to avoid the moisture's condense, so the MacEtch reaction temperature is 55 °C and the reaction is considered to happen with a solid–gas interface.

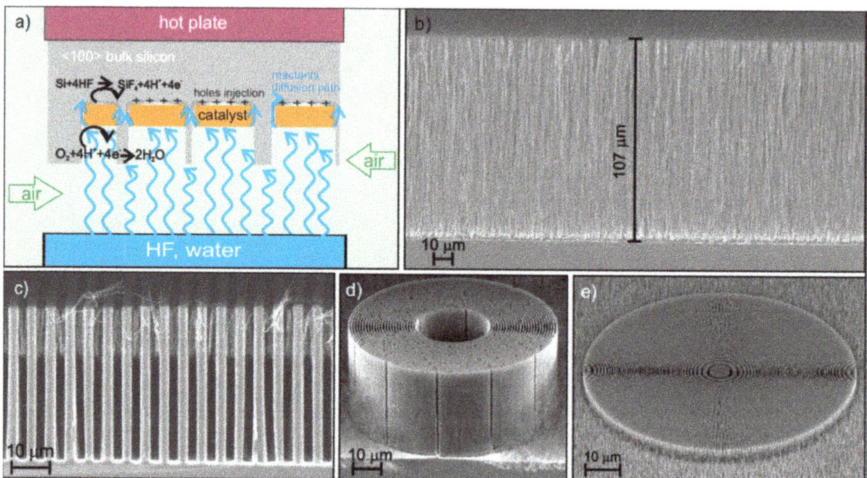

Figure 10. (**a**) Schematic of MacEtch in gas phase. SEM in cross section of structures by gas phase MacEtch at 55 °C, HF was evaporated from a water diluted HF solution and the oxidant is supplied by air: (**b**) Si nanowires; (**c**) linear grating with pitch of 4.8 µm; (**d**) circular grating with pitch 1 µm; (**e**) zone plate with outmost pitch of 200 nm. The figure was adapted with permission from L. Romano, 2020. [37].

The use of gas-MacEtch turned out to be very useful to improve the stability of free-standing Si nanostructures such as the nanowires and the zone plate in the outmost region. A totally interconnected catalyst design results in free-standing Si nanopillars. As noted by R. Akan et al. [35] for very high aspect ratios and smallest zone sizes, these pillars will become mechanically unstable. Chang et al. [17] and K. Li et al. [36] increased the number of Si interconnects to solve this issue, but this further reduces the active zone plate area and consequently the efficiency. The zone plate made by gas-MacEtch (Figure 10e) does not need Si interconnects since the stability of Si lamellas is not compromised by the liquid drying.

Moreover, nanowires can be used as diffractive optics in speckle based X-ray phase contrast imaging [79]. Nanowires are expected to improve the sensitivity by producing speckles of smaller size in comparison to sandpaper [80] or other membranes with feature size in the micrometer scale.

4. Silicon Based Microfabrication of X-ray Optics

Microstructures constitute the X-ray optical elements such as diffractive and refractive X-ray lenses for microfocusing applications at synchrotron beam lines [81] and diffractive gratings for interferometric systems [9,82]. For low energy applications (<10 keV) microfabrication can be realized in silicon but higher X-ray absorbing materials are necessary for hard X-rays. The most common microfabrication approach is based on creating a low X-ray absorbing template and filling it with highly absorbing metal. Historically, the templates are based either on polymer [83,84] or silicon [85].

Silicon microfabrication has been the key technology in manufacturing integrated circuits and microchips in the semiconductor industry. This gives the advantage of a well-assessed technology with a competitive mass production such as deep reactive ion etching [86] and KOH wet etching [87,88]. The combination of unconventional processing and the freedom from microelectronics constrains enrich the spectrum of capabilities and give a new life to the "old silicon material", with revolutionizing advancements in nanotechnology [37]. We recently reported about micro- and nano-fabrication processing for X-ray gratings, including lithography [89], dry [37,90,91] and wet etching methods [39], Au electroplating [22], Ir atomic layer deposition [92] and metal casting [19,20]. The use of MacEtch as a microfabrication process for X-ray optical devices was first reported in 2014 for Fresnel zone plate structures [17,18]. Some SEM images of Fresnel zone plate structures produced by MacEtch are reported in Figure 3; Figure 10. The X-ray nanofocusing effect of Fresnel zone plate optics fabricated by MacEtch and atomic layer deposition of Pt was recently reported by K. Li et al. [93]. In the following section we report an example of X-ray grating interferometry with gratings fabricated by MacEtch. The main challenge for X-ray grating interferometry is the fabrication of the absorption gratings [85], which are metal periodic microstructures, for high energy X-ray (>30 keV).

Deep X-ray lithography (also called LIGA) [83,84] is used to pattern the polymer template. This technology has the advantage that the polymer pattern can be created on whatever substrate, such as a metallic substrate that is used as a seed layer for the following Au electroplating process in order to create the final Au absorbing grating. The metal layer can be deposited on graphite that has the advantage of being flexible and allows to easily bend the Au grating structure. However, LIGA process is limited to relatively small area (10×10 cm^2) and it is quite expensive since it requires a synchrotron facility. In the case of Si based technology, the Si etched structure can affect the quality of the Au electroplating filling, some distortions [22] or voiding inside the template, resulting in less X-ray diffraction efficiency. Atomic layer deposition of metallic coating has been implemented to create a metallization layer for Au electroplating with conventional damascene approaches [94,95] and Au bottom up superfilling processes [96]. MacEtch offers the possibility to benefit of the original catalyst layer as a seed for the Au electroplating filling.

Gratings Fabrication for X-ray Phase-Contrast Imaging

X-ray grating interferometry (GI) based imaging is a very promising, fast growing and competitive technique for medical, material science and security applications [1]. Contrast in X-ray imaging with GI can be boosted by exploiting refraction and scattering, in addition to conventional absorption. GI might have a large impact on the radiological approach to medical X-ray imaging because it will intrinsically enable the detection of subtle differences in the electron density of a material (like a lesion delineation) and the measurement of the effective integrated local small-angle scattering power generated by the microscopic structural fluctuations in the specimen (such as micro-calcifications in a breast tissue for instance). Similar enhancements are expected in homeland security or material science application, where structural properties such as orientation, degree of anisotropy, average structure size, and distribution of structural sizes can be inferred via omnidirectional tensor tomography [97].

The purpose of an X-ray interferometer is to encode propagation-induced phase changes in the beam wavefront—when passing through a specimen—into an intensity modulation measured by a (usually position sensitive) detector placed downstream. In its simplest configuration, called Talbot Interferometer, an X-ray interferometer consists of two gratings placed in a partially coherent beam.

The latter is usually provided by a third/fourth generation synchrotron source or, with significantly less intensity, by a microfocus X-ray tube, see Figure 11. The first grating G_1 (of period p_1) is usually a phase-grating, i.e., it actually does not absorb the beam but it imposes a significant phase shift resulting in a controlled wavefront modulation at a specific distance downstream, usually where the second, absorbing grating G_2 (of period p_2) is placed. G_1 essentially divides the incoming beam into the two first diffraction orders: being the grating pitch ($p_1 \sim \mu m$) much larger than the incoming wavelength ($\sim \mathring{A}$), the resulting angle between both diffracted beams is so small that they almost fully overlap, resulting in a linear periodic interference fringe pattern downstream of G_1, in planes perpendicular to the optical axis. This effect is known as the fractional Talbot effect [98]. A sample of interest is placed either in front or behind G_1, and it usually absorbs, refracts and scatters the incoming beam. These interactions consequently affect the interference pattern: absorption leads to an average intensity reduction, refraction causes a lateral displacement of the fringes and scattering reduces the fringe amplitude. For a phase grating with a phase shift of π illuminated by a plane wave, the periodicity of the fringe pattern equals $p_1/2$ [9]. The detector resolution might not be good enough to resolve the interference pattern and therefore a second, absorbing grating G_2 (with the same periodicity as the fringes) is placed immediately in front of the detector at the position where the fringes form. This grating behaves as a transmission (analyzer) mask and converts local fringe positions into signal intensity variations. This is a crucial aspect of GI, as the analyzer gratings *de-facto* decouples the phase sensitivity of the system from its intrinsic spatial resolution, making GI suitable for operation on large samples and large field of views. In fact, when the source does not provide a sufficiently high spatial coherence, like in the case of a conventional X-ray tube, a third grating G_0 of period p_0 can be introduced right after the source yielding to the so-called Talbot-Lau (Figure 11a) configuration. G_0 is an absorbing structure that creates an array of individually coherent, but mutually incoherent sources. If the condition $p_0 = p_2 \times L_{01}/L_{12}$ is fulfilled [85], where L_{01} is the distance between G_0 and G_1 and L_{12} is distance between G_1 and G_2, then the images created by each line source are superimposed in the image plane. This enables to carry out efficient phase contrast X-ray imaging on commercially, normally incoherent sources. Retrieval of the absorption, phase, and scattering signals has been done with various methods, with the phase-stepping [9] and the fringe scanning [99,100] being the most common used approaches.

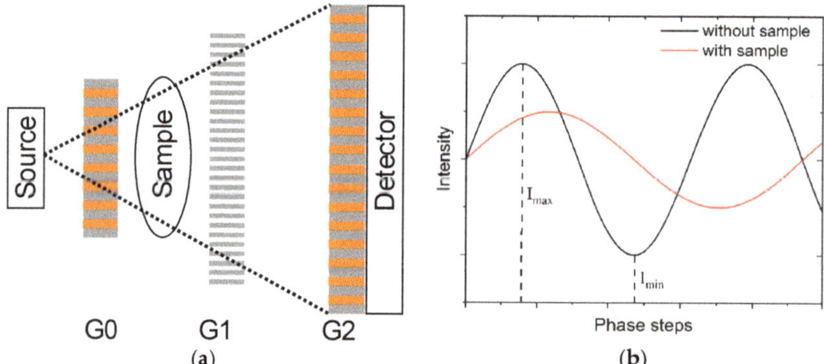

Figure 11. (**a**) Sketch of an X-ray grating interferometer in the Talbot–Lau configuration. (**b**) Scheme of the phase stepping process.

The intensity modulation of the recorded fringe pattern is usually characterized by its visibility (Equation (5)):

$$V = \frac{I_{max} - I_{min}}{I_{max} + I_{min}} \tag{5}$$

where *Imax* and *Imin* are indicated in the phase step curve of Figure 11b. The visibility depends on the degree of spatial coherence of the illumination as well as its spectrum, on the system geometry and on grating's pitch and depth. The interference fringe visibility is a common figure of merit for the design of X-ray gratings interferometers [101]. This is because the formation of high-modulation fringe pattern is a prerequisite for robust grating interferometry.

The main challenge is the fabrication of the absorption gratings [85], which are metal periodic microstructures with high aspect ratio that are usually fabricated starting from templates produced by LIGA [84] or deep Si etching [22,39,85]. The period p_2 (p_0) of the absorbing grating G_2 (G_0) is usually in the range of 1–20 µm (5–100 µm), while the height (h) depends on X-ray energy and absorption efficiency of the material [102]. A transmission of the structures of less than 25% is acceptable. It can be calculated [102] that 10 µm thickness is sufficient for photon energies below 20 keV, while for photon energies of 30 keV (60 keV), about 25 µm (160 µm) of gold is required. Very challenging, medically oriented projects, require the design of very sensitive interferometers extending on short geometries, imposing quite extreme aspect-ratios for G_2 for which periods p_2 as small as 1 µm (or below) and gold height of 30 µm might be needed [103]. Such requirements pushed the research efforts toward MacEtch as a new technique that is able to provide aspect ratio structures with period ranging from tens of nanometers to tens of micrometers. Moreover, MacEtch offers the possibility to benefit of the original catalyst layer as a seed for the Au electroplating filling. Figure 12 summarizes the Si template fabrication by using MacEtch and the subsequent Au electroplating.

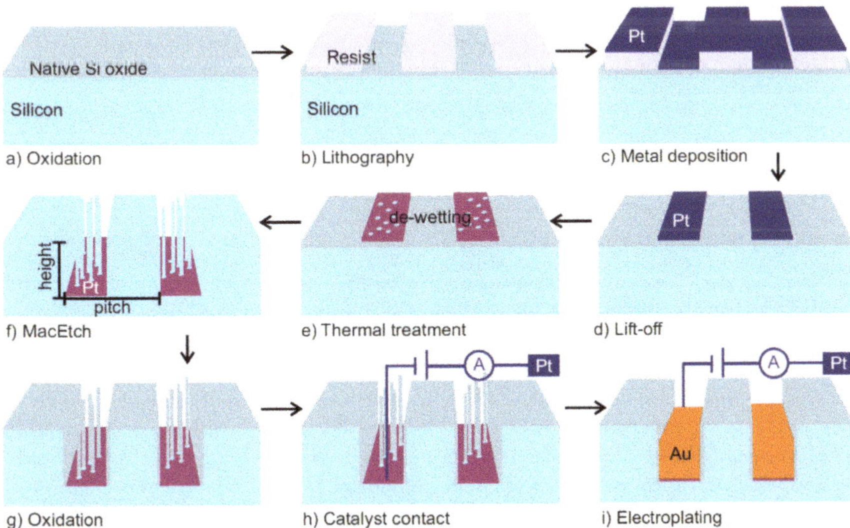

Figure 12. Schematic illustration of the grating fabrication process by MacEtch and subsequent Au electroplating: (**a**) growth of native silicon oxide on Si substrate, (**b**) pattern definition by means of a lithographic process, (**c**) Pt deposition by evaporation, (**d**) lift-off, (**e**) Pt de-wetting by thermal treatment, (**f**) MacEtch in a solution of HF and H_2O_2, (**g**) Si side wall oxidation in air, (**h**) electrical contact of the catalyst metal interconnected pattern to the electroplating electrode, (**i**) seeded Au growth by electroplating. The Pt pattern works as a catalyst for MacEtch and as a seed layer for Au electroplating. The figure was adapted with permission from L. Romano et al., 2020. [21].

Absorption gratings are usually fabricated by metal electroplating (typically of Au, which is one of the most efficient absorbing materials for X-rays), into high aspect ratio Si templates. The performance of the Au electroplated grating in terms of uniformity and quality of the filling can be well assessed with an X-ray interferometric set up. Figure 13 reports a typical example of grating characterization. The performance of an Au filled grating with a pitch of 6 μm was investigated with an X-ray interferometer [104] (design energy 20 keV) operated at the 3rd Talbot order with π/2 phase shifting G1 grating. The Pt-MacEtch grating (pitch 6 μm, height 39 μm) filled with Au up to 30 μm was used as G0 grating together with a G1 phase grating made of Si by deep reactive ion etching (DRIE) and a G2 absorbing grating fabricated by DRIE and Au electroplating [22]. The average X-ray fringe visibility of 17.5% (Figure 13a) is comparable to the values achieved in absorbing gratings fabricated by conventional DRIE followed by Au electroplating [22]. Figure 13b,c is examples of images obtained with the X-ray grating interferometer, phase contrast (Figure 13b) enhances the detection of low absorbing textures, while dark field (Figure 13c) highlights the presence of microstructures.

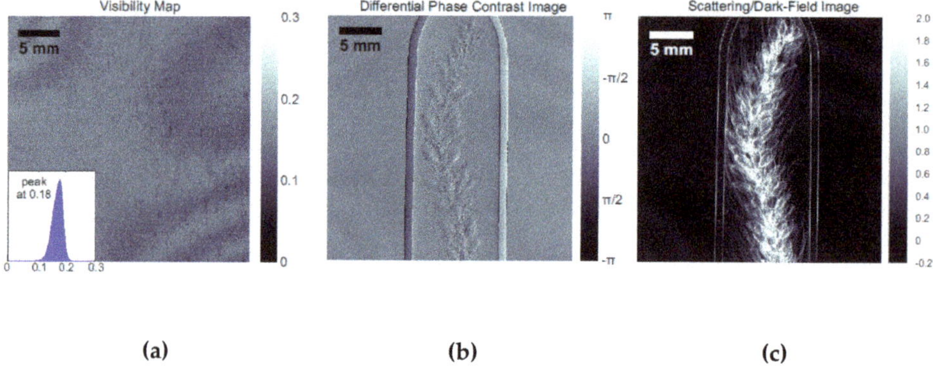

Figure 13. X-ray performance of an Au electroplated grating of 6 μm period used as a G$_0$ absorbing grating. X-ray fringe visibility map and visibility histogram as insert (**a**), differential phase contrast (**b**), and scattering (**c**) images of a grain ear. The figure was adapted with permission from L. Romano et al., 2020. [21].

5. Conclusions and Perspectives

MacEtch is a very powerful and promising technique that is competing the performances of more conventional etching technology, such as deep reactive ion etching and cryogenic processes [86]. Having a clear idea about the fundamentals and recent advances in this area allows researchers to have a better perspective. During the process of etching, a triggering event that produces unequal etch rates might occur. These events can change the effective forces on the catalyst and produce a resultant torque on the catalyst [51]. Such triggering events can deteriorate the etching uniformity and compromise the quality of the pattern transfer from the original lithography to the Si substrate. We discussed the role of the metal catalyst pattern in order to be able to extent the feature size of the etched structure from nanometers to micrometers and to control the etching uniformity. Metal de-wetting technique turned out to be a reliable method to create a porous catalyst layer that allows to etch very high aspect ratio structures. We discussed the critical role of etching solution by showing the effect of reactants depletion and the presence of alcohol additives. A small amount of alcohol can help to improve the etching uniformity and reduce the formation of the microporosity on the side wall trenches. MacEtch in gas phase showed new possibilities of etching nanostructures with exceptional aspect ratio up to 10,000:1. In the framework of X-ray optics, the fabrication of useful microstructures by MacEtch with successful story of implementations started to record new publications in the last few years, with continuously improving performances. In order to give an idea, we selected all the publications about MacEtch that are relevant to produce high aspect ratio structures in silicon (100) substrates for lithographic patterns such as ordered pillars arrays, electron beam lithographic patterns, gratings, Micro Electro Mechanical Systems (MEMS) microstructures, photonic crystals, etc. Figure 14 reports the number of publications as a function of the publication year to the best of the authors' knowledge. The total number of publications in the time range of 2008–2020 years is 81, including this special issue of Micromachines, the trend indicates a substantial increment in the last 10 years. The publications addressing the use of MacEtch for specific X-ray optics fabrication is 13 with the first reports in 2014.

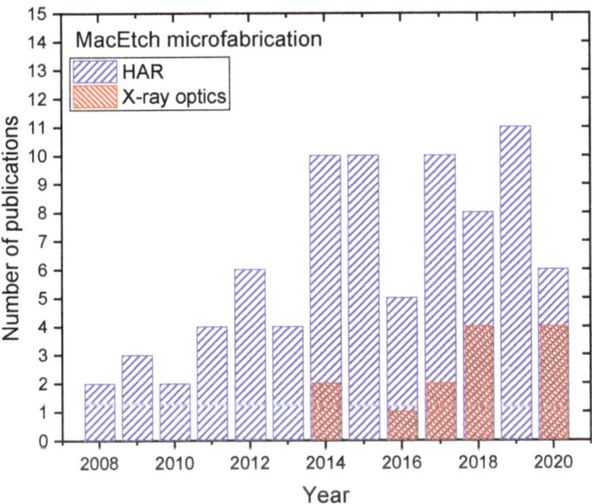

Figure 14. Number of publications as a function of year regarding MacEtch for producing high aspect ratio (HAR) patterned structures [2,8,15,17,18,21,24–26,31,32,35–41,43,47,50,51,54,56–58,62,63,71,73,77, 78,93,105–153] and specific for X-ray optics fabrication [17,18,21,35–41,93,119,126,153].

X-ray optics with nanostructured features, such as kinoform lenses and Fresnel zone plates, showed to benefit from MacEtch fabrication. We can envisage new future applications of gas-phase MacEtch when high aspect ratio and feature size in the nanoscale are needed, such in X-ray microscopy with energy higher than 20 keV. A full process of gratings microfabrication for X-ray interferometry by using MacEtch and subsequent Au electroplating is demonstrated with performances similar to other fabrication methods. The Pt catalyst layer that sinks down into the Si substrate during MacEtch has been successfully used as seed layer for Au electroplating in order to fabricate a periodic structure with an high absorbing material for hard X-ray. The possibility to push forward the aspect ratio and the relative low fabrication costs on large area of MacEtch with respect to other technologies, such as reactive ion etching and LIGA, are motivating the investigation efforts. We envisage that MacEtch will become a new enabling technology to fulfill the requirements of grating's height for absorbing hard X-ray radiation and submicron grating's pitch for boosting the sensitivity in grating based high sensitive X-ray systems, as required for instance for early breast cancer detection.

The main bottleneck in grating based X-ray interferometry is the fabrication of high aspect ratio periodic structures, whose quality and homogeneity over large areas strongly affect the contrast of the generated images. There is the need to produce X-ray diffraction gratings with (i) very high aspect ratio (AR $\geq$ 50:1) in highly absorbing material such as gold; (ii) large area (mammography, e.g., asks for a field of view above 20 × 20 cm^2 [103,154]); (iii) good uniformity (no distortions and changes in the duty cycle and depth over the grating area); and (iv) bending capability in order to improve the field of view limitation of cone beam emission from the X-ray source. MacEtch technique has the clear advantage of silicon patterning with high aspect ratio at nanoscale, it is a relatively low cost technology since it is accessible even in labs with limited equipment (no vacuum or clean-room conditions). However, MacEtch is performed in solutions of hydrofluoric acid, we recommend to follow all the necessary safety protocols in order to handle heavily concentrated hydrofluoric acid. The bending capability of the silicon substrate depends on the wafer thickness, extreme bending has been reported for silicon wafers with thickness below 50 µm but the bending of a thick absorbing grating can be challenging and we predict that dedicated processing needs to be developed to avoid cracks and distortions, such as eventually etching off the silicon template. Four inch wafer scale MacEtch gratings have been demonstrated with good uniformity and high control of trench profile and etching direction for aspect ratio up to 30:1 [39]. The process itself has no limits in terms of patterning area such as LIGA, photolithographic processes are available for Si based technology up to 12 inch wafer scale, periodic linear gratings with pitch size in the range of 100 nm can be patterned on wafer scale by interference lithography [89], nanoimprinting processes can be implemented to further increase the patterned area with nanoscale resolution [55]. High quality structures with high aspect ratio require an etching regime that is dominated by the diffusion of the reactive species. Stirring [26] and large volume of solutions [21] are used to homogenize the concentration of reactants during the etching but the effect on etching rate, aspect ratio and defects need to be systematically investigated in order to anticipate the commercialization of MacEtch as a grating fabrication technology. The fabrication throughput as a function of grating quality and performances needs to be assessed to start the technology transfer from the research laboratory level to an industrial R&D.

MacEtch as a technology is still at its infancy, its control and reproducibility over a large area are still not clear and n to be systematically studied. Further scientific efforts need to be made to take full advantage of high aspect ratio capability and exploit its application.

Author Contributions: Both authors contributed equally. All authors have read and agreed to the published version of the manuscript.

Funding: We acknowledge the support from: SNF Sinergia Grant CRSII5_18356 "Clinical GI-BCT", EU Grant ERC-2012-StG 310005 "PhaseX", ERC-2016-PoC 727246 "Magic", Eurostar Grant E!1106 "INFORMAT" and NanoArgovia Grant 13.01 "NANOCREATE" (Swiss Nanoscience Institute) and Fondazione Gelu (TACI-C Project).

Acknowledgments: We would like to thank K. Jefimovs, J. Vila-Comamala, M. Kagias (PSI-TOMCAT) and C. David (PSI-LMN) for useful discussion.

Conflicts of Interest: The authors declare no conflict of interest.

References

1. Sze, S.M.; Ng, K.K. *Physics of Semiconductor Devices*, 3rd ed.; John Wiley & Sons Inc.: Hoboken, NJ, USA, 2006.
2. Zahedinejad, M.; Farimani, S.D.; Khaje, M.; Mehrara, H.; Erfanian, A.; Zeinali, F. Deep and vertical silicon bulk micromachining using metal assisted chemical etching. *J. Micromech. Microeng.* **2013**, *23*, 055015. [CrossRef]
3. Hamelin, B.; Li, L.; Daruwalla, A.; Wong, C.P.; Ayazi, F. In High-aspect-ratio sub-micron trench etching on SOI using wet metal-assisted chemical etching (MaCE) process. In Proceedings of the IEEE 29th International Conference on Micro Electro Mechanical Systems (MEMS), Shanghai, China, 24–28 January 2016; pp. 447–450. [CrossRef]
4. Sze, S.M. *Semiconductor Sensors*; John Wiley & Sons Inc.: Hoboken, NJ, USA, 1994.
5. Snyder, G.J.; Lim, J.R.; Huang, C.-K.; Fleurial, J.-P. Thermoelectric microdevice fabricated by a MEMS-like electrochemical process. *Nat. Mater.* **2003**, *2*, 528–531. [CrossRef] [PubMed]
6. Zamfir, M.R.; Nguyen, H.T.; Moyen, E.; Lee, Y.H.; Pribat, D. Silicon nanowires for Li-based battery anodes: A review. *J. Mater. Chem. A* **2013**, *1*, 9566–9586. [CrossRef]
7. Garnett, E.; Yang, P. Light Trapping in Silicon Nanowire Solar Cells. *Nano Lett.* **2010**, *10*, 1082–1087. [CrossRef] [PubMed]
8. Balasundaram, K.; Mohseni, P.K.; Shuai, Y.-C.; Zhao, D.; Zhou, W.; Li, X. Photonic crystal membrane reflectors by magnetic field-guided metal-assisted chemical etching. *Appl. Phys. Lett.* **2013**, *103*, 214103. [CrossRef]
9. Weitkamp, T.; Diaz, A.; David, C.; Pfeiffer, F.; Stampanoni, M.; Cloetens, P.; Ziegler, E. X-ray phase imaging with a grating interferometer. *Opt. Express* **2005**, *13*, 6296–6304. [CrossRef] [PubMed]
10. Wu, B.; Kumar, A.; Pamarthy, S. High aspect ratio silicon etch: A review. *J. Appl. Phys.* **2010**, *108*, 051101. [CrossRef]

11. Mondiali, V.; Lodari, M.; Chrastina, D.; Barget, M.; Bonera, E.; Bollani, M. Micro and nanofabrication of SiGe/Ge bridges and membranes by wet-anisotropic etching. *Microelectron. Eng.* **2015**, *141*, 256–260. [CrossRef]

12. Sato, K.; Shikida, M.; Matsushima, Y.; Yamashiro, T.; Asaumi, K.; Iriye, Y.; Yamamoto, M. Characterization of orientation-dependent etching properties of single-crystal silicon: Effects of KOH concentration. *Sens. Actuators A Phys.* **1998**, *64*, 87–93. [CrossRef]

13. Li, X.; Bohn, P.W. Metal-assisted chemical etching in HF/H2O2 produces porous silicon. *Appl. Phys. Lett.* **2000**, *77*, 2572–2574. [CrossRef]

14. Han-Don, U.; Namwoo, K.; Kangmin, L.; Inchan, H.; Ji, H.S.; Young, J.Y.; Peter, D.; Munib, W.; Kwanyong, S. Versatile control of metal-assisted chemical etching for vertical silicon microwire arrays and their photovoltaic applications. *Sci. Rep.* **2015**, *5*, 11277.

15. Huang, Z.; Geyer, N.; Werner, P.; de Boor, J.; Gösele, U. Metal-Assisted Chemical Etching of Silicon: A Review. *Adv. Mater.* **2011**, *2*, 285–308. [CrossRef] [PubMed]

16. Li, X. In MacEtch: Anisotropic metal-assisted chemical etching defies the textbooks. *SPIE Newsroom* **2012**. [CrossRef]

17. Chang, C.; Sakdinawat, A. Ultra-high aspect ratio high-resolution nanofabrication for hard X-ray diffractive optics. *Nat. Commun* **2014**, *5*, 4243. [CrossRef] [PubMed]

18. Tiberio, R.C.; Rooks, M.J.; Chang, C.; Knollenberg, C.F.; Dobisz, E.A.; Sakdinawat, A. Vertical directionality-controlled metal-assisted chemical etching for ultrahigh aspect ratio nanoscale structures. *J. Vac. Sci. Technol. Bnanotechnol. Microelectron. Mater. Process. Meas. Phenom.* **2014**, *32*, 06FI01. [CrossRef]

19. Romano, L.; Vila-Comamala, J.; Kagias, M.; Vogelsang, K.; Schift, H.; Stampanoni, M.; Jefimovs, K. High aspect ratio metal microcasting by hot embossing for X-ray optics fabrication. *Microelectron. Eng.* **2017**, *176*, 6–10. [CrossRef]

20. Romano, L.; Vila-Comamala, J.; Schift, H.; Stampanoni, M.; Jefimovs, K. Hot embossing of Au- and Pb-based alloys for x-ray grating fabrication. *J. Vac. Sci. Technol. Bnanotechnol. Microelectron. Mater. Process. Meas. Phenom.* **2017**, *35*, 06G302. [CrossRef]

21. Romano, L.; Vila-Comamala, J.; Jefimovs, K.; Stampanoni, M. High aspect ratio gratings microfabrication by Pt assisted chemical etching and Au electroplating. *Adv. Eng. Mater.* **2020**, in press. [CrossRef]

22. Kagias, M.; Wang, Z.; Guzenko, V.A.; David, C.; Stampanoni, M.; Jefimovs, K. Fabrication of Au gratings by seedless electroplating for X-ray grating interferometry. *Mater. Sci. Semicond. Process.* **2019**, *92*, 73–79. [CrossRef]

23. Hildreth, O.; Wong, C.P. Nano-metal-Assisted Chemical Etching for Fabricating Semiconductor and Optoelectronic Devices. In *Materials for Advanced Packaging*; Lu, D., Wong, C.P., Eds.; Springer International Publishing: Cham, Switzerland, 2017; pp. 879–922.

24. Huang, Z.; Zhang, X.; Reiche, M.; Liu, L.; Lee, W.; Shimizu, T.; Senz, S.; Gösele, U. Extended Arrays of Vertically Aligned Sub-10 nm Diameter [100] Si Nanowires by Metal-Assisted Chemical Etching. *Nano Lett.* **2008**, *8*, 3046–3051. [CrossRef]

25. Hildreth, O.J.; Brown, D.; Wong, C.P. 3D Out-of-Plane Rotational Etching with Pinned Catalysts in Metal-Assisted Chemical Etching of Silicon. *Adv. Funct. Mater.* **2011**, *21*, 3119–3128. [CrossRef]

26. Li, L.; Zhang, G.; Wong, C.P. Formation of Through Silicon Vias for Silicon Interposer in Wafer Level by Metal-Assisted Chemical Etching. *IEEE Trans. Compon. Packag. Manuf. Technol.* **2015**, *5*, 1039–1049. [CrossRef]

27. Huang, H.-C.; Kim, M.; Zhan, X.; Chabak, K.; Kim, J.D.; Kvit, A.; Liu, D.; Ma, Z.; Zuo, J.-M.; Li, X. High Aspect Ratio β-Ga2O3 Fin Arrays with Low-Interface Charge Density by Inverse Metal-Assisted Chemical Etching. *ACS Nano* **2019**, *13*, 8784–8792. [CrossRef] [PubMed]

28. Yeo, C.; Kim, J.B.; Song, Y.M.; Lee, Y.T. Antireflective silicon nanostructures with hydrophobicity by metal-assisted chemical etching for solar cell applications. *Nanoscale Res. Lett.* **2013**, *8*, 159. [CrossRef] [PubMed]

29. Li, X. Metal assisted chemical etching for high aspect ratio nanostructures: A review of characteristics and applications in photovoltaics. *Curr. Opin. Solid State Mater. Sci.* **2012**, *16*, 71–81. [CrossRef]

30. Kim, M.; Yi, S.; Kim, J.D.; Liu, S.; Zhou, W.; Yu, Z.; Li, X. In Nano-indented Ge surfaces by metal-assisted chemical etching (MacEtch) and its application for optoelectronic device. In Proceedings of the 75th Annual Device Research Conference (DRC), South Bend, IN, USA, 25–28 June 2017; pp. 1–2.

31. Van Toan, N.; Toda, M.; Hokama, T.; Ono, T. Cantilever with High Aspect Ratio Nanopillars on Its Top Surface for Moisture Detection in Electronic Products. *Adv. Eng. Mater.* **2017**, *19*, 1700203. [CrossRef]

32. Li, L.; Zhao, X.; Wong, C.-P. Deep Etching of Single- and Polycrystalline Silicon with High Speed, High Aspect Ratio, High Uniformity, and 3D Complexity by Electric Bias-Attenuated Metal-Assisted Chemical Etching (EMaCE). *ACS Appl. Mater. Interfaces* **2014**, *6*, 16782–16791. [CrossRef]

33. DeJarld, M.; Shin, J.C.; Chern, W.; Chanda, D.; Balasundaram, K.; Rogers, J.A.; Li, X. Formation of High Aspect Ratio GaAs Nanostructures with Metal-Assisted Chemical Etching. *Nano Lett.* **2011**, *11*, 5259–5263. [CrossRef]

34. Chen, Y.; Zhang, C.; Li, L.; Zhou, S.; Chen, X.; Gao, J.; Zhao, N.; Wong, C.-P. Hybrid Anodic and Metal-Assisted Chemical Etching Method Enabling Fabrication of Silicon Carbide Nanowires. *Small* **2019**, *15*, 1803898. [CrossRef]

35. Akan, R.; Parfeniukas, K.; Vogt, C.; Toprak, M.S.; Vogt, U. Reaction control of metal-assisted chemical etching for silicon-based zone plate nanostructures. *RSC Adv.* **2018**, *8*, 12628–12634. [CrossRef]

36. Li, K.; Wojcik, M.J.; Divan, R.; Ocola, L.E.; Shi, B.; Rosenmann, D.; Jacobsen, C. Fabrication of hard x-ray zone plates with high aspect ratio using metal-assisted chemical etching. *J. Vac. Sci. Technol. B Nanotechnol. Microelectron. Mater. Process. Meas. Phenom.* **2017**, *35*, 06G901. [CrossRef]

37. Romano, L.; Kagias, M.; Vila-Comamala, J.; Jefimovs, K.; Tseng, L.-T.; Guzenko, V.A.; Stampanoni, M. Metal assisted chemical etching of silicon in the gas phase: A nanofabrication platform for X-ray optics. *Nanoscale Horiz.* **2020**, *5*, 869–879. [CrossRef]

38. Akan, R.; Parfeniukas, K.; Vogt, C.; Toprak, M.S.; Vogt, U. Investigation of Metal-Assisted Chemical Etching for Fabrication of Silicon-Based X-ray Zone Plates. *Microsc. Microanal.* **2018**, *24*, 286–287. [CrossRef]

39. Romano, L.; Kagias, M.; Jefimovs, K.; Stampanoni, M. Self-assembly nanostructured gold for high aspect ratio silicon microstructures by metal assisted chemical etching. *RSC Adv.* **2016**, *6*, 16025–16029. [CrossRef]

40. Romano, L.; Vila-Comamala, J.; Jefimovs, K.; Stampanoni, M. Effect of isopropanol on gold assisted chemical etching of silicon microstructures. *Microelectron. Eng.* **2017**, *177*, 59–65. [CrossRef]

41. Lebugle, M.; Dworkowski, F.; Pauluhn, A.; Guzenko, V.A.; Romano, L.; Meier, N.; Marschall, F.; Sanchez, D.F.; Grolimund, D.; Wang, M.; et al. High-intensity x-ray microbeam for macromolecular crystallography using silicon kinoform diffractive lenses. *Appl. Opt.* **2018**, *57*, 9032–9039. [CrossRef] [PubMed]

42. Kolasinski, K.W. The mechanism of galvanic/metal-assisted etching of silicon. *Nanoscale Res. Lett.* **2014**, *9*, 432. [CrossRef]

43. Chiappini, C. MACE Silicon Nanostructures. In *Handbook of Porous Silicon*; Canham, L., Ed.; Springer International Publishing: Cham, Switzerland, 2017; pp. 12–21.

44. Lévy-Clément, C. Porous Silicon Formation by Metal Nanoparticle-Assisted Etching. In *Handbook of Porous Silicon*; Canham, L., Ed.; Springer International Publishing: Cham, Switzerland, 2014.

45. Hu, Y.; Peng, K.-Q.; Qiao, Z.; Huang, X.; Zhang, F.-Q.; Sun, R.-N.; Meng, X.-M.; Lee, S.-T. Metal-Catalyzed Electroless Etching of Silicon in Aerated HF/H2O Vapor for Facile Fabrication of Silicon Nanostructures. *Nano Lett.* **2014**, *14*, 4212–4219. [CrossRef]

46. Chartier, C.; Bastide, S.; Lévy-Clément, C. Metal-assisted chemical etching of silicon in HF–H2O2. *Electrochim. Acta* **2008**, *53*, 5509–5516. [CrossRef]

47. Hildreth, O.J.; Lin, W.; Wong, C.P. Effect of Catalyst Shape and Etchant Composition on Etching Direction in Metal-Assisted Chemical Etching of Silicon to Fabricate 3D Nanostructures. *ACS Nano* **2009**, *3*, 4033–4042. [CrossRef] [PubMed]

48. Huang, Z.; Shimizu, T.; Senz, S.; Zhang, Z.; Geyer, N.; Gösele, U. Oxidation Rate Effect on the Direction of Metal-Assisted Chemical and Electrochemical Etching of Silicon. *J. Phys. Chem. C* **2010**, *114*, 10683–10690. [CrossRef]

49. Lehmann, V. *The Electrochemistry of Silicon: Instrumentation, Science, Materials and Applications*; Wiley-VCH: Weinheim, Germany, 2002.

50. Kim, J.; Han, H.; Kim, Y.H.; Choi, S.-H.; Kim, J.-C.; Lee, W. Au/Ag Bilayered Metal Mesh as a Si Etching Catalyst for Controlled Fabrication of Si Nanowires. *ACS Nano* **2011**, *5*, 3222–3229. [CrossRef] [PubMed]

51. Kong, L.; Zhao, Y.; Dasgupta, B.; Ren, Y.; Hippalgaonkar, K.; Li, X.; Chim, W.K.; Chiam, S.Y. Minimizing Isolate Catalyst Motion in Metal-Assisted Chemical Etching for Deep Trenching of Silicon Nanohole Array. *ACS Appl. Mater. Interfaces* **2017**, *9*, 20981–20990. [CrossRef] [PubMed]

52. Hildreth, O.J.; Rykaczewski, K.; Fedorov, A.G.; Wong, C.P. A DLVO model for catalyst motion in metal-assisted chemical etching based upon controlled out-of-plane rotational etching and force-displacement measurements. *Nanoscale* **2013**, *5*, 961–970. [CrossRef] [PubMed]

53. Li, L.; Zhao, X.; Wong, C.-P. Charge Transport in Uniform Metal-Assisted Chemical Etching for 3D High-Aspect-Ratio Micro- and Nanofabrication on Silicon. *ECS J. Solid State Sci. Technol.* **2015**, *4*, P337–P346. [CrossRef]

54. Rykaczewski, K.; Hildreth, O.J.; Kulkarni, D.; Henry, M.R.; Kim, S.-K.; Wong, C.P.; Tsukruk, V.V.; Fedorov, A.G. Maskless and Resist-Free Rapid Prototyping of Three-Dimensional Structures Through Electron Beam Induced Deposition (EBID) of Carbon in Combination with Metal-Assisted Chemical Etching (MaCE) of Silicon. *ACS Appl. Mater. Interfaces* **2010**, *2*, 969–973. [CrossRef] [PubMed]

55. Li, L.; Li, B.; Zhang, C.; Tuan, C.-C.; Lin, Z.; Wong, C.-P. A facile and low-cost route to high-aspect-ratio microstructures on silicon via a judicious combination of flow-enabled self-assembly and metal-assisted chemical etching. *J. Mater. Chem. C* **2016**, *4*, 8953–8961. [CrossRef]

56. Oh, Y.; Choi, C.; Hong, D.; Kong, S.D.; Jin, S. Magnetically Guided Nano–Micro Shaping and Slicing of Silicon. *Nano Lett.* **2012**, *12*, 2045–2050. [CrossRef]

57. Hildreth, O.J.; Fedorov, A.G.; Wong, C.P. 3D Spirals with Controlled Chirality Fabricated Using Metal-Assisted Chemical Etching of Silicon. *ACS Nano* **2012**, *6*, 10004–10012. [CrossRef]

58. Han, H.; Huang, Z.; Lee, W. Metal-assisted chemical etching of silicon and nanotechnology applications. *Nano Today* **2014**, *9*, 271–304. [CrossRef]

59. Peng, K.Q.; Hu, J.J.; Yan, Y.J.; Wu, Y.; Fang, H.; Xu, Y.; Lee, S.T.; Zhu, J. Fabrication of Single-Crystalline Silicon Nanowires by Scratching a Silicon Surface with Catalytic Metal Particles. *Adv. Funct. Mater.* **2006**, *16*, 387–394. [CrossRef]

60. Mikhael, B.; Elise, B.; Xavier, M.; Sebastian, S.; Johann, M.; Laetitia, P. New Silicon Architectures by Gold-Assisted Chemical Etching. *ACS Appl. Mater. Interfaces* **2011**, *3*, 3866–3873. [CrossRef] [PubMed]

61. Liu, X.; Coxon, P.R.; Peters, M.; Hoex, B.; Cole, J.M.; Fray, D.J. Black silicon: Fabrication methods, properties and solar energy applications. *Energy Environ. Sci.* **2014**, *7*, 3223–3263. [CrossRef]

62. Lianto, P.; Yu, S.; Wu, J.; Thompson, C.V.; Choi, W.K. Vertical etching with isolated catalysts in metal-assisted chemical etching of silicon. *Nanoscale* **2012**, *4*, 7532–7539. [CrossRef] [PubMed]

63. Li, L.; Liu, Y.; Zhao, X.; Lin, Z.; Wong, C.-P. Uniform Vertical Trench Etching on Silicon with High Aspect Ratio by Metal-Assisted Chemical Etching Using Nanoporous Catalysts. *ACS Appl. Mater. Interfaces* **2014**, *6*, 575–584. [CrossRef]

64. Yae, S.; Morii, Y.; Fukumuro, N.; Matsuda, H. Catalytic activity of noble metals for metal-assisted chemical etching of silicon. *Nanoscale Res. Lett.* **2012**, *7*, 352. [CrossRef]

65. Lai, C.Q.; Cheng, H.; Choi, W.K.; Thompson, C.V. Mechanics of Catalyst Motion during Metal Assisted Chemical Etching of Silicon. *J. Phys. Chem. C* **2013**, *117*, 20802–20809. [CrossRef]

66. Li, L.; Tuan, C.; Zhang, C.; Chen, Y.; Lian, G.; Wong, C. Uniform Metal-Assisted Chemical Etching for Ultra-High-Aspect-Ratio Microstructures on Silicon. *J. Microelectromech. Syst.* **2018**, *28*, 143–153. [CrossRef]

67. Kim, J.D.; Kim, M.; Kong, L.; Mohseni, P.K.; Ranganathan, S.; Pachamuthu, J.; Chim, W.K.; Chiam, S.Y.; Coleman, J.J.; Li, X. Self-Anchored Catalyst Interface Enables Ordered Via Array Formation from Submicrometer to Millimeter Scale for Polycrystalline and Single-Crystalline Silicon. *ACS Appl. Mater. Interfaces* **2018**, *10*, 9116–9122. [CrossRef]

68. Thompson, C.V. Solid-State Dewetting of Thin Films. *Annu. Rev. Mater. Res.* **2012**, *42*, 399–434. [CrossRef]

69. Strobel, S.; Kirkendall, C.; Chang, J.B.; Berggren, K.K. Sub-10 nm structures on silicon by thermal dewetting of platinum. *Nanotechnology* **2010**, *21*, 505301. [CrossRef] [PubMed]

70. Rola, K.P.; Zubel, I. Impact of alcohol additives concentration on etch rate and surface morphology of (100) and (110) Si substrates etched in KOH solutions. *Microsyst. Technol.* **2013**, *19*, 635–643. [CrossRef]

71. Li, Y.; Duan, C. Bubble-Regulated Silicon Nanowire Synthesis on Micro-Structured Surfaces by Metal-Assisted Chemical Etching. *Langmuir* **2015**, *31*, 12291–12299. [CrossRef] [PubMed]

72. Cassie, A.B.D.; Baxter, S. Wettability of porous surfaces. *Trans. Faraday Soc.* **1944**, *40*, 546–551. [CrossRef]

73. Balasundaram, K.; Jyothi, S.S.; Jae Cheol, S.; Bruno, A.; Debashis, C.; Mohammad, M.; Keng, H.; John, A.R.; Placid, F.; Sanjiv, S.; et al. Porosity control in metal-assisted chemical etching of degenerately doped silicon nanowires. *Nanotechnology* **2012**, *23*, 305304. [CrossRef]

74. Kim, Y.; Tsao, A.; Lee, D.H.; Maboudian, R. Solvent-induced formation of unidirectionally curved and tilted Si nanowires during metal-assisted chemical etching. *J. Mater. Chem. C* **2013**, *1*, 220–224. [CrossRef]

75. Cheng, S.L.; Chung, C.H.; Lee, H.C. A Study of the Synthesis, Characterization, and Kinetics of Vertical Silicon Nanowire Arrays on (001)Si Substrates. *J. Electrochem. Soc.* **2008**, *155*, D711–D714. [CrossRef]

76. Yan, J.; Wu, S.; Zhai, X.; Gao, X.; Li, X. Facile fabrication of wafer-scale, micro-spacing and high-aspect-ratio silicon microwire arrays. *Rsc Adv.* **2016**, *6*, 87486–87492. [CrossRef]

77. Li, H.; Ye, T.; Shi, L.; Xie, C. Fabrication of ultra-high aspect ratio (>160:1) silicon nanostructures by using Au metal assisted chemical etching. *J. Micromech. Microeng.* **2017**, *27*, 124002. [CrossRef]

78. Hildreth, O.J.; Schmidt, D.R. Vapor Phase Metal-Assisted Chemical Etching of Silicon. *Adv. Funct. Mater.* **2014**, *24*, 3827–3833. [CrossRef]

79. Bérujon, S.; Ziegler, E.; Cerbino, R.; Peverini, L. Two-Dimensional X-ray Beam Phase Sensing. *Phys. Rev. Lett.* **2012**, *10*, 158102. [CrossRef]

80. Wang, H.; Kashyap, Y.; Sawhney, K. From synchrotron radiation to lab source: Advanced speckle-based X-ray imaging using abrasive paper. *Sci. Rep.* **2016**, *6*, 20476. [CrossRef] [PubMed]

81. David, C.; Weitkamp, T.; Nöhammer, B.; van der Veen, J.F. Diffractive and refractive X-ray optics for microanalysis applications. *Spectrochim. Acta Part. B At. Spectrosc.* **2004**, *59*, 1505–1510. [CrossRef]

82. Rutishauser, S.; Bednarzik, M.; Zanette, I.; Weitkamp, T.; Börner, M.; Mohr, J.; David, C. Fabrication of two-dimensional hard X-ray diffraction gratings. *Microelectron. Eng.* **2013**, *101*, 12–16. [CrossRef]

83. Noda, D.; Tanaka, M.; Shimada, K.; Yashiro, W.; Momose, A.; Hattori, T. Fabrication of large area diffraction grating using LIGA process. *Microsyst. Technol.* **2008**, *14*, 1311–1315. [CrossRef]

84. Mohr, J.; Grund, T.; Kunka, D.; Kenntner, J.; Leuthold, J.; Meiser, J.; Schulz, J.; Walter, M. High aspect ratio gratings for X-ray phase contrast imaging. *Aip Conf. Proc.* **2012**, *1466*, 41–50.

85. David, C.; Bruder, J.; Rohbeck, T.; Grünzweig, C.; Kottler, C.; Diaz, A.; Bunk, O.; Pfeiffer, F. Fabrication of diffraction gratings for hard X-ray phase contrast imaging. *Microelectron. Eng.* **2007**, *84*, 1172–1177. [CrossRef]

86. Ishikawa, K.; Karahashi, K.; Ishijima, T.; Cho, S.I.; Elliott, S.; Hausmann, D.; Mocuta, D.; Wilson, A.; Kinoshita, K. Progress in nanoscale dry processes for fabrication of high-aspect-ratio features: How can we control critical dimension uniformity at the bottom? *Jpn. J. Appl. Phys.* **2018**, *57*, 06JA01. [CrossRef]

87. Finnegan, P.S.; Hollowell, A.E.; Arrington, C.L.; Dagel, A.L. High aspect ratio anisotropic silicon etching for x-ray phase contrast imaging grating fabrication. *Mater. Sci. Semicond. Process.* **2019**, *92*, 80–85. [CrossRef]

88. Hollowell, A.E.; Arrington, C.L.; Finnegan, P.; Musick, K.; Resnick, P.; Volk, S.; Dagel, A.L. Double sided grating fabrication for high energy X-ray phase contrast imaging. *Mater. Sci. Semicond. Process.* **2019**, *92*, 86–90. [CrossRef]

89. Jefimovs, K.; Romano, L.; Vila-Comamala, J.; Kagias, M.E.; Wang, Z.; Wang, L.; Dais, C.; Solak, H.; Stampanoni, M. High-aspect ratio silicon structures by displacement Talbot lithography and Bosch etching. In Proceedings of the Advances in Patterning Materials and Processes XXXIV, San Jose, CA, USA, 27 March 2017; SPIE: San Jose, CA, USA, 2017; p. 101460L.

90. Kagias, M.; Wang, Z.; Jefimovs, K.; Stampanoni, M. Dual phase grating interferometer for tunable dark-field sensitivity. *Appl. Phys. Lett.* **2017**, *110*, 014105. [CrossRef]

91. Kagias, M.; Wang, Z.; Villanueva-Perez, P.; Jefimovs, K.; Stampanoni, M. 2D-Omnidirectional Hard-X-ray Scattering Sensitivity in a Single Shot. *Phys. Rev. Lett.* **2016**, *116*, 093902. [CrossRef] [PubMed]

92. Vila-Comamala, J.; Romano, L.; Guzenko, V.; Kagias, M.; Stampanoni, M.; Jefimovs, K. Towards sub-micrometer high aspect ratio X-ray gratings by atomic layer deposition of iridium. *Microelectron. Eng.* **2018**, *192*, 19–24. [CrossRef]

93. Li, K.; Ali, S.; Wojcik, M.; De Andrade, V.; Huang, X.; Yan, H.; Chu, Y.S.; Nazaretski, E.; Pattammattel, A.; Jacobsen, C. Tunable hard x-ray nanofocusing with Fresnel zone plates fabricated using deep etching. *Optica* **2020**, *7*, 410–416. [CrossRef]

94. Znati, S.; Chedid, N.; Miao, H.; Chen, L.; Bennett, E.E.; Wen, H. Electrodeposition of Gold to Conformally Fill High-Aspect-Ratio Nanometric Silicon Grating Trenches: A Comparison of Pulsed and Direct Current Protocols. *J. Surf. Eng. Mater. Adv. Technol.* **2015**, *5*, 7. [CrossRef] [PubMed]

95. Song, T.-E.; Lee, S.; Han, H.; Jung, S.; Kim, S.-H.; Kim, M.J.; Lee, S.W.; Ahn, C.W. Evaluation of grating realized via pulse current electroplating combined with atomic layer deposition as an x-ray grating interferometer. *J. Vac. Sci. Technol. A* **2019**, *37*, 030903. [CrossRef]

96. Josell, D.; Ambrozik, S.; Williams, M.E.; Hollowell, A.E.; Arrington, C.; Muramoto, S.; Moffat, T.P. Exploring the Limits of Bottom-Up Gold Filling to Fabricate Diffraction Gratings. *J. Electrochem. Soc.* **2019**, *166*, D898–D907. [CrossRef]

97. Kim, J.; Kagias, M.; Marone, F.; Stampanoni, M. X-ray scattering tensor tomography with circular gratings. *Appl. Phys. Lett.* **2020**, *116*, 134102. [CrossRef]

98. Talbot, H.F. LXXVI. Facts relating to optical science. No. IV. *Lond. Edinb. Dublin Philos. Mag. J. Sci.* **1836**, *9*, 401–407. [CrossRef]

99. Arboleda, C.; Wang, Z.; Stampanoni, M. Tilted-grating approach for scanning-mode X-ray phase contrast imaging. *Opt. Express* **2014**, *22*, 15447–15458. [CrossRef] [PubMed]

100. Kottler, C.; Pfeiffer, F.; Bunk, O.; Grünzweig, C.; David, C. Grating interferometer based scanning setup for hard x-ray phase contrast imaging. *Rev. Sci. Instrum.* **2007**, *78*, 043710. [CrossRef] [PubMed]

101. Yan, A.; Wu, X.; Liu, H. Predicting visibility of interference fringes in X-ray grating interferometry. *Opt. Express* **2016**, *24*, 15927–15939. [CrossRef] [PubMed]

102. Henke, B.L.; Gullikson, E.M.; Davis, J.C. X-ray Interactions: Photoabsorption, Scattering, Transmission, and Reflection at E = 50–30,000 eV, Z = 1–92. *At. Data Nucl. Data Tables* **1993**, *54*, 181–342. [CrossRef]

103. Arboleda, C.; Wang, Z.; Koehler, T.; Martens, G.; Van Stevendaal, U.; Bartels, M.; Villanueva-Perez, P.; Roessl, E.; Stampanoni, M. Sensitivity-based optimization for the design of a grating interferometer for clinical X-ray phase contrast mammography. *Opt. Express* **2017**, *25*, 6349–6364. [CrossRef] [PubMed]

104. Donath, T.; Chabior, M.; Pfeiffer, F.; Bunk, O.; Reznikova, E.; Mohr, J.; Hempel, E.; Popescu, S.; Hoheisel, M.; Schuster, M.; et al. Inverse geometry for grating-based x-ray phase-contrast imaging. *J. Appl. Phys.* **2009**, *106*, 054703. [CrossRef]

105. Kim, T.K.; Bae, J.-H.; Kim, J.; Kim, Y.-C.; Jin, S.; Chun, D.W. Bulk Micromachining of Si by Annealing-Driven Magnetically Guided Metal-Assisted Chemical Etching. *ACS Appl. Electron. Mater.* **2020**, *2*, 260–267. [CrossRef]

106. Chien, P.-J.; Wei, T.-C.; Chen, C.-Y. High-Speed and Direction-Controlled Formation of Silicon Nanowire Arrays Assisted by Electric Field. *Nanoscale Res. Lett.* **2020**, *15*, 25. [CrossRef]

107. Zhang, X.; Hu, Y.; Gu, H.; Zhu, P.; Jiang, W.; Zhang, G.; Sun, R.; Wong, C.-P. A Highly Sensitive and Cost-Effective Flexible Pressure Sensor with Micropillar Arrays Fabricated by Novel Metal-Assisted Chemical Etching for Wearable Electronics. *Adv. Mater. Technol.* **2019**, *4*, 1900367. [CrossRef]

108. Zarei, S.; Zahedinejad, M.; Mohajerzadeh, S. Metal-assisted chemical etching for realisation of deep silicon microstructures. *Micro Nano Lett.* **2019**, *14*, 1083–1086. [CrossRef]

109. Wilhelm, T.S.; Kecskes, I.L.; Baboli, M.A.; Abrand, A.; Pierce, M.S.; Landi, B.J.; Puchades, I.; Mohseni, P.K. Ordered Si Micropillar Arrays via Carbon-Nanotube-Assisted Chemical Etching for Applications Requiring Nonreflective Embedded Contacts. *ACS Appl. Nano Mater.* **2019**, *2*, 7819–7826. [CrossRef]

110. Wang, S.; Liu, H.; Han, J. Comprehensive Study of Au Nano-Mesh as a Catalyst in the Fabrication of Silicon Nanowires Arrays by Metal-Assisted Chemical Etching. *Coatings* **2019**, *9*, 149. [CrossRef]

111. Van Toan, N.; Wang, X.; Inomata, N.; Toda, M.; Voiculescu, I.; Ono, T. Low Cost and High-Aspect Ratio Micro/Nano Device Fabrication by Using Innovative Metal-Assisted Chemical Etching Method. *Adv. Eng. Mater.* **2019**, *21*, 1900490. [CrossRef]

112. Shimizu, T.; Niwa, R.; Ito, T.; Shingubara, S. Effect of additives on preparation of vertical holes in Si substrate using metal-assisted chemical etching. *Jpn. J. Appl. Phys.* **2019**, *58*, SDDF06. [CrossRef]

113. Pérez-Díaz, O.; Quiroga-González, E.; Silva-González, N.R. Silicon microstructures through the production of silicon nanowires by metal-assisted chemical etching, used as sacrificial material. *J. Mater. Sci.* **2019**, *54*, 2351–2357. [CrossRef]

114. Obata, S.; Sano, M.; Shimokawa, K.; Higuchi, K. A Novel Fabrication Process for High Density Silicon Capacitors by using Metal-Assisted Chemical Etching. *Int. Symp. Microelectron.* **2019**, *2019*, 248–253. [CrossRef]

115. Kim, J.D.; Kim, M.; Chan, C.; Draeger, N.; Coleman, J.J.; Li, X. CMOS-Compatible Catalyst for MacEtch: Titanium Nitride-Assisted Chemical Etching in Vapor phase for High Aspect Ratio Silicon Nanostructures. *ACS Appl. Mater. Interfaces* **2019**, *11*, 27371–27377. [CrossRef] [PubMed]

116. Hsin, C.; Wu, M.; Wang, W. Thermoelectric Devices by Half-Millimeter-Long Silicon Nanowires Arrays. *IEEE Trans. Nanotechnol.* **2019**, *18*, 921–924. [CrossRef]

117. Baytemir, G.; Ciftpinar, E.H.; Turan, R. Enhanced metal assisted etching method for high aspect ratio microstructures: Applications in silicon micropillar array solar cells. *Sol. Energy* **2019**, *194*, 148–155. [CrossRef]

118. Xu, J.; Liu, G.; Huang, Q.; Liu, M.; Zhou, X.; Wu, H.; Li, N.; Li, Y.; Xu, X.; Liang, D. Kinoform and saw-tooth X-ray refractive lenses development at SSRF. *J. Instrum.* **2018**, *13*, C07005. [CrossRef]

119. Parfeniukas, K. *High-Aspect Ratio Nanofabrication for Hard X-ray Zone Plates*; KTH Royal Institute of Technology: Stockholm, Sweden, 2018.

120. Moldovan, N.; Divan, R.; Zeng, H.; Ocola, L.E.; De Andrade, V.; Wojcik, M. Atomic layer deposition frequency-multiplied Fresnel zone plates for hard x-rays focusing. *J. Vac. Sci. Technol. A Vac. Surf. Film.* **2018**, *36*, 01A124. [CrossRef]

121. Li, X.; Kim, J.D.; Kim, M.; Kong, L. Self-Anchored Catalyst Metal-Assisted Chemical Etching. U.S. Patent 10134599B2, 20 November 2018.

122. Kumar, J.; Ingole, S. Effect of Silicon Conductivity and HF/H2O2 Ratio on Morphology of Silicon Nanostructures Obtained via Metal-Assisted Chemical Etching. *J. Electron. Mater.* **2018**, *47*, 1583–1588. [CrossRef]

123. Zhang, J.; Zhang, L.; Han, L.; Tian, Z.-W.; Tian, Z.-Q.; Zhan, D. Electrochemical nanoimprint lithography: When nanoimprint lithography meets metal assisted chemical etching. *Nanoscale* **2017**, *9*, 7476–7482. [CrossRef] [PubMed]

124. Teng, F.; Li, N.; Xu, D.; Xiao, D.; Yang, X.; Lu, N. Precise regulation of tilt angle of Si nanostructures via metal-assisted chemical etching. *Nanoscale* **2017**, *9*, 449–453. [CrossRef] [PubMed]

125. Parfeniukas, K.; Giakoumidis, S.; Akan, R.; Vogt, U. High-aspect ratio zone plate fabrication for hard x-ray nanoimaging. In Proceedings of the Advances in X-ray/EUV Optics and Components XII, International Society for Optics and Photonics, San Diego, CA, USA, 23 August 2017; p. 103860S.

126. Lebugle, M.; Seniutinas, G.; Marschall, F.; Guzenko, V.A.; Grolimund, D.; David, C. Tunable kinoform x-ray beam splitter. *Opt. Lett.* **2017**, *42*, 4327–4330. [CrossRef] [PubMed]

127. Kim, J.D.; Mohseni, P.K.; Balasundaram, K.; Ranganathan, S.; Pachamuthu, J.; Coleman, J.J.; Li, X. Scaling the Aspect Ratio of Nanoscale Closely Packed Silicon Vias by MacEtch: Kinetics of Carrier Generation and Mass Transport. *Adv. Funct. Mater.* **2017**, *27*, 1605614. [CrossRef]

128. Cozzi, C.; Polito, G.; Kolasinski, K.W.; Barillaro, G. Controlled Microfabrication of High-Aspect-Ratio Structures in Silicon at the Highest Etching Rates: The Role of H2O2 in the Anodic Dissolution of Silicon in Acidic Electrolytes. *Adv. Funct. Mater.* **2017**, *27*, 9. [CrossRef]

129. Wu, R.W.; Yuan, G.D.; Wang, K.C.; Wei, T.B.; Liu, Z.Q.; Wang, G.H.; Wang, J.X.; Li, J.M. Bilayer–metal assisted chemical etching of silicon microwire arrays for photovoltaic applications. *Aip Adv.* **2016**, *6*, 025324. [CrossRef]

130. Smith, B.D.; Patil, J.J.; Ferralis, N.; Grossman, J.C. Catalyst Self-Assembly for Scalable Patterning of Sub 10 nm Ultrahigh Aspect Ratio Nanopores in Silicon. *ACS Appl. Mater. Interfaces* **2016**, *8*, 8043–8049. [CrossRef]

131. Li, L. Uniform high-aspect-ratio 3D micro-and nanomanufacturing on silicon by (electro)-metal-assisted chemical etching: Fundamentals and applications. Ph.D. Thesis, Georgia Institute of Technology, Atlanta, GA, USA, August 2016.

132. Lai, R.A.; Hymel, T.M.; Narasimhan, V.K.; Cui, Y. Schottky Barrier Catalysis Mechanism in Metal-Assisted Chemical Etching of Silicon. *ACS Appl. Mater. Interfaces* **2016**, *8*, 8875–8879. [CrossRef]

133. Wang, Y.M.; Lu, L.X.; Srinivasan, B.M.; Asbahi, M.; Zhang, Y.W.; Yang, J.K.W. High aspect ratio 10-nm-scale nanoaperture arrays with template-guided metal dewetting. *Sci. Rep.* **2015**, *5*, 7. [CrossRef]

134. Li, L.; Holmes, C.M.; Hah, J.; Hildreth, O.J.; Wong, C.P. Uniform Metal-assisted Chemical Etching and the Stability of Catalysts. *MRS Proc.* **2015**, *1801*, 1–8. [CrossRef]

135. Lai, C.Q.; Zheng, W.; Choi, W.K.; Thompson, C.V. Metal assisted anodic etching of silicon. *Nanoscale* **2015**, *7*, 11123–11134. [CrossRef] [PubMed]

136. Chiappini, C.; De Rosa, E.; Martinez, J.O.; Liu, X.; Steele, J.; Stevens, M.M.; Tasciotti, E. Biodegradable silicon nanoneedles delivering nucleic acids intracellularly induce localized in vivo neovascularization. *Nat. Mater.* **2015**, *14*, 532. [CrossRef] [PubMed]

137. Chiappini, C.; Campagnolo, P.; Almeida, C.S.; Abbassi-Ghadi, N.; Chow, L.W.; Hanna, G.B.; Stevens, M.M. Mapping Local Cytosolic Enzymatic Activity in Human Esophageal Mucosa with Porous Silicon Nanoneedles. *Adv. Mater.* **2015**, *27*, 5147–5152. [CrossRef] [PubMed]

138. Chen, C.-Y.; Liu, Y.-R.; Tseng, J.-C.; Hsu, P.-Y. Uniform trench arrays with controllable tilted profiles using metal-assisted chemical etching. *Appl. Surf. Sci.* **2015**, *333*, 152–156. [CrossRef]

139. Yeom, J.; Ratchford, D.; Field, C.R.; Brintlinger, T.H.; Pehrsson, P.E. Decoupling Diameter and Pitch in Silicon Nanowire Arrays Made by Metal-Assisted Chemical Etching. *Adv. Funct. Mater.* **2014**, *24*, 106–116. [CrossRef]

140. Lin, H.; Fang, M.; Cheung, H.-Y.; Xiu, F.; Yip, S.; Wong, C.-Y.; Ho, J.C. Hierarchical silicon nanostructured arrays via metal-assisted chemical etching. *Rsc Adv.* **2014**, *4*, 50081–50085. [CrossRef]

141. Lai, C.Q.; Choi, W.K. Synthesis of free-standing, curved Si nanowires through mechanical failure of a catalyst during metal assisted chemical etching. *Phys. Chem. Chem. Phys.* **2014**, *16*, 13402–13408. [CrossRef]

142. Booker, K.; Brauers, M.; Crisp, E.; Rahman, S.; Weber, K.; Stocks, M.; Blakers, A. Metal-assisted chemical etching for very high aspect ratio grooves inn-type silicon wafers. *J. Micromech. Microeng.* **2014**, *24*, 125026. [CrossRef]

143. Wong, C.; Li, L.; Hildreth, O. Nano etching via metal-assisted chemical etching (mace) for through silicon via (tsv) stacked chips application. In Proceedings of the Advanced Optoelectronics for Energy and Environment, Wuhan, China, 25–26 May 2013; Optical Society of America: Washington, DC, USA, 2013; p. 3.

144. Hildreth, O.J.; Honrao, C.; Sundaram, V.; Wong, C.P. Combining Electroless Filling with Metal-Assisted Chemical Etching to Fabricate 3D Metallic Structures with Nanoscale Resolutions. *Ecs Solid State Lett.* **2013**, *2*, P39–P41. [CrossRef]

145. Shin, J.C.; Chanda, D.; Chern, W.; Yu, K.J.; Rogers, J.A.; Li, X. Experimental Study of Design Parameters in Silicon Micropillar Array Solar Cells Produced by Soft Lithography and Metal-Assisted Chemical Etching. *IEEE J. Photovolt.* **2012**, *2*, 129–133. [CrossRef]

146. Geyer, N.; Fuhrmann, B.; Huang, Z.; de Boor, J.; Leipner, H.S.; Werner, P. Model for the Mass Transport during Metal-Assisted Chemical Etching with Contiguous Metal Films as Catalysts. *J. Phys. Chem. C* **2012**, *116*, 13446–13451. [CrossRef]

147. Lee, D.H.; Kim, Y.; Doerk, G.S.; Laboriante, I.; Maboudian, R. Strategies for controlling Si nanowire formation during Au-assisted electroless etching. *J. Mater. Chem.* **2011**, *21*, 10359–10363. [CrossRef]

148. Chang, S.-W.; Oh, J.; Boles, S.T.; Thompson, C.V. Fabrication of silicon nanopillar-based nanocapacitor arrays. *Appl. Phys. Lett.* **2010**, *96*, 153108. [CrossRef]

149. Huang, Z.; Shimizu, T.; Senz, S.; Zhang, Z.; Zhang, X.; Lee, W.; Geyer, N.; Gösele, U. Ordered Arrays of Vertically Aligned [110] Silicon Nanowires by Suppressing the Crystallographically Preferred <100> Etching Directions. *Nano Lett.* **2009**, *9*, 2519–2525. [CrossRef]

150. Chang, S.-W.; Chuang, V.P.; Boles, S.T.; Ross, C.A.; Thompson, C.V. Densely Packed Arrays of Ultra-High-Aspect-Ratio Silicon Nanowires Fabricated using Block-Copolymer Lithography and Metal-Assisted Etching. *Adv. Funct. Mater.* **2009**, *19*, 2495–2500. [CrossRef]

151. Choi, W.K.; Liew, T.H.; Dawood, M.K.; Smith, H.I.; Thompson, C.V.; Hong, M.H. Synthesis of Silicon Nanowires and Nanofin Arrays Using Interference Lithography and Catalytic Etching. *Nano Lett.* **2008**, *8*, 3799–3802. [CrossRef]

152. Ki, B.; Song, Y.; Choi, K.; Yum, J.H.; Oh, J. Chemical Imprinting of Crystalline Silicon with Catalytic Metal Stamp in Etch Bath. *ACS Nano* **2018**, *12*, 609–616. [CrossRef]

153. Akan, R.; Frisk, T.; Lundberg, F.; Ohlin, H.; Johansson, U.; Li, K.; Sakdinawat, A.; Vogt, U. Metal-Assisted Chemical Etching and Electroless Deposition for Fabrication of Hard X-ray Pd/Si Zone Plates. *Micromachines* **2020**, *11*, 301. [CrossRef]

154. Vila-Comamala, J.; Arboleda, C.; Romano, L.; Kuo, W.; Lang, K.; Jefimovs, K.; Wang, Z.; Singer, G.; Vine, D.; Yun, W.; et al. Development of Laboratory Grating-based X-ray Phase Contrast Microtomography for Improved Pathology. *Microsc. Microanal.* **2018**, *24*, 190–191. [CrossRef]

© 2020 by the authors. Licensee MDPI, Basel, Switzerland. This article is an open access article distributed under the terms and conditions of the Creative Commons Attribution (CC BY) license (http://creativecommons.org/licenses/by/4.0/).

MDPI

St. Alban-Anlage 66

4052 Basel

Switzerland

Tel. +41 61 683 77 34

Fax +41 61 302 89 18

www.mdpi.com

Micromachines Editorial Office

E-mail: micromachines@mdpi.com

www.mdpi.com/journal/micromachines

www.ingramcontent.com/pod-product-compliance
Lightning Source LLC
LaVergne TN
LVHW070618170726
843515LV00004B/118